Lecture Notes in Networks and Systems 768

The series "Lecture Notes in Networks and Systems" publishes the latest developments in Networks and Systems—quickly, informally and with high quality. Original research reported in proceedings and post-proceedings represents the core of LNNS.

Volumes published in LNNS embrace all aspects and subfields of, as well as new challenges in, Networks and Systems.

The series contains proceedings and edited volumes in systems and networks, spanning the areas of Cyber-Physical Systems, Autonomous Systems, Sensor Networks, Control Systems, Energy Systems, Automotive Systems, Biological Systems, Vehicular Networking and Connected Vehicles, Aerospace Systems, Automation, Manufacturing, Smart Grids, Nonlinear Systems, Power Systems, Robotics, Social Systems, Economic Systems and other. Of particular value to both the contributors and the readership are the short publication timeframe and the world-wide distribution and exposure which enable both a wide and rapid dissemination of research output.

The series covers the theory, applications, and perspectives on the state of the art and future developments relevant to systems and networks, decision making, control, complex processes and related areas, as embedded in the fields of interdisciplinary and applied sciences, engineering, computer science, physics, economics, social, and life sciences, as well as the paradigms and methodologies behind them.

Indexed by SCOPUS, INSPEC, WTI Frankfurt eG, zbMATH, SCImago.

All books published in the series are submitted for consideration in Web of Science.

For proposals from Asia please contact Aninda Bose (aninda.bose@springer.com).

Muhammad Younas · Irfan Awan ·
Salima Benbernou · Dana Petcu
Editors

The 4th Joint International Conference on Deep Learning, Big Data and Blockchain (DBB 2023)

Springer

Editors
Muhammad Younas
School of Engineering, Computing and Mathematics
Oxford Brookes University
Oxford, UK

Irfan Awan
Department of Computer Science, Faculty of Engineering and Informatics
University of Bradford
Bradford, UK

Salima Benbernou
Université Paris Cité
Paris, France

Dana Petcu
Computer Science Department, Faculty of Mathematics and Informatics
Western University of Timisoara
Timisoara, Romania

ISSN 2367-3370 ISSN 2367-3389 (electronic)
Lecture Notes in Networks and Systems
ISBN 978-3-031-42316-1 ISBN 978-3-031-42317-8 (eBook)
https://doi.org/10.1007/978-3-031-42317-8

This Springer imprint is published by the registered company Springer Nature Switzerland AG
The registered company address is: Gewerbestrasse 11, 6330 Cham, Switzerland

Preface

It was a great pleasure to welcome all the participants of the 4th Joint International Conference on Deep Learning, Big Data and Blockchain (DBB 2023). The conference was held during August 14–16, 2023, both online and onsite, in the historic city of Marrakech, Morocco. Marrakech is one of the world-famous cities, which attracts millions of visitors from all over the world. The country name Morocco stems itself from Marrakech. The city has historical landmarks such as museums, squares, castles, gardens and other modern attractions such as shopping areas and restaurants.

The DBB 2023 conference involved hard work, time and commitment from the conference organizing and technical committees. The goal was to allow participants to share and exchange ideas on different topics related to the conference's theme, including machine/deep learning, blockchain, big data and their integration in modern applications and convergence in new and emerging research and development areas.

The call for papers of the conference included innovative and timely topics in the aforementioned areas and their sub-topics, such as learning-based models; clustering, classification and regression; data analysis, insights and hidden patterns; blockchain protocols and applications; verification; security and trust; and applications of deep learning, blockchain and big data in areas such as business, finance and healthcare among others. Blockchain and smart contract tools and methods are increasingly used in new and emerging systems to ensure transparency of data and transactions. For instance, machine learning techniques have been used by businesses to analyze a large volume of (big) data and to identify useful patterns in the data so that they can be used for purposes of intelligent and timely decision-making.

The conference technical committee created a fascinating technical program to provide a forum where participants could present, discuss and provide constructive feedback on different aspects of deep learning, big data and blockchain.

The conference has attracted good-quality papers from different countries worldwide. The conference followed a rigorous review process wherein multiple technical program committee members reviewed all submitted papers. Based on the reviews, eleven papers were accepted for the conference, which gave an acceptance rate of 33% of the total submissions.

The accepted papers included interesting work on timely and emerging research topics such as distributed ledger technology and environment; blockchains in the healthcare domain; blockchain architectures; interoperability of blockchain systems; neural network application; machine learning in disease prediction; deep learning applications; smart cities; machine learning in flight pricing; and energy demand. The papers also included practical work in application domains such as healthcare and commercial systems.

We sincerely thank all the members of the program committee who have spent their valuable time reviewing the submitted papers and providing useful feedback to the authors. We were also thankful to all the authors for their contributions to the conference.

We were grateful to the members of the conference committee for their support: Dana Petcu and Abdeslam En-Nouaary (General Co-chairs), Ismail Berrada and Loubna Mekouar (Local Organizing Co-chairs), Filipe Portela (Workshop Coordinator), Betül Ay (Publicity Chair) and Natalia Kryvinska (Journal Special Issue Coordinator).

We sincerely thank Springer's team for the time and support they provided throughout the production of the conference proceedings.

August 2023

Salima Benbernou
Muhammad Younas

Organization

DBB 2023 Organizing Committee

General Co-chairs

Dana Petcu	West University of Timisoara, Romania
Abdeslam En-Nouaary	INPT, Morocco

Program Co-chairs

Salima Benbernou	Université Paris Cité, France
Muhammad Younas	Oxford Brookes University, UK

Publication Chair

Irfan Awan	University of Bradford, UK

Local Organizing Co-chairs

Ismail Berrada	UM6P, Morocco
Loubna Mekouar	UM6P, Morocco

Journal Special Issue Coordinator

Natalia Kryvinska	Comenius University in Bratislava, Slovakia

Workshop Coordinator

Filipe Portela	University of Minho Portugal, Portugal

Publicity Chair

Betül Ay	Fırat University, Turkey

Program Committee

Ahmed Ratnani	UM6P, Morocco
Angelika Kedzierska-Szczepaniak	University of Gdansk, Poland
Antonio Dourado	University of Coimbra, Portugal
Boris Kovalerchuk	Central Washington University, USA
Bruno Veloso	INESC Technology and Science, Porto, Portugal
Changze Cui	Amazon Web Services, USA
Chirine Ghedira Guegan	Université Lyon 3, France
Chouki Tibermacine	Université de Montpellier, France
Christian Eitzinger	Profactor GmbH, Steyr-Gleink, Austria
Daniela Zaharie	West University of Timisoara, Romania
Daniele Apiletti	Polytechnic University of Turin, Italy
Darell Long	University of California, Santa Cruz, USA
Felix J. Garcia Clemente	University of Murcia, Spain
Grigorios Beligiannis	University of Patras, Greece
Gulden Kokturk	Dokuz Eylul University, Turkey
Hamed Taherdoost	Hamta Business Corporation, Canada
Hang Guo	University of Southern California, USA
Hassina Meziane	Université of Oran, Algeria
Huiru (Jane) Zheng	Ulster University, UK
Karim El Moutaouakil	University Sidi Mohamed Ben Abdellah, Morocco
Lei Zhang	East China Normal University, China
Nelson Jorge Ribeiro Duarte	Porto Polytechnic, Poland
Nizar Bouguila	Concordia University, Canada
Noor Akhmad Setiawan	Universitas Gadjah Mada, Indonesia
Pavas Navaney	Oracle America Inc., USA
Pavel Loskot	ZJU-UIUC Institute, China
Qiang Cheng	University of Kentucky, USA
Rabiah Ahmad	Universiti Teknikal Malaysia, Malaysia
Sergi Trilles Oliver	Universitat Jaume I, Spain
Shahid Ali	Manukau Institute of Technology, New Zealand
Sotiris Kotsiantis	University of Patras, Greece
Sung-Bae Cho	Yonsei University, Korea
Tomoyuki Uchida	Hiroshima City University, Japan
Valentina Emilia Balas	Aurel Vlaicu University of Arad, Romania
Vijay Anant Athavale	Walchand Institute of Technology, India
Xuan Guo	University of North Texas, USA
Yiming Xu	Amazon, Washington, USA
Yinpu Li	Florida State University, USA
Zografoula Vagena	Université de Paris, France

Contents

Block Chain Systems

Distributed Ledger Technology for Collective Environmental Action

Roman Beck[1](✉), Marco Schletz[2], Alvise Baggio[3], and Lorenzo Gentile[1]

[1] IT University of Copenhagen, 2300 Copenhagen, Denmark
beck@itu.dk

[2] Department of Public Policy, University of North Carolina at Chapel Hill, Chapel Hill, NC, USA

[3] Politecnico Di Milano, 20133 Milano, Italy

Abstract. The Paris Agreement sets forth a global effort to limit climate warming to well below 2 °C, necessitating collaborative actions among economically competitive nations. This research introduces a distributed ledger technology (DLT)-based system designed to uphold sovereign data control while facilitating cross-national enforcement of CO_2 emissions monitoring and reduction policies. Drawing insights from the implementation of vehicular fuel consumption metering in the European Union, we leverage the framework of coopetition theory to elucidate the potential alignment of disparate interests towards a common environmental goal, with due regard to national autonomy. We present a DLT prototype, developed via a design science methodology, aimed at monitoring and reducing automotive CO_2 emissions across Europe. This prototype serves as an exemplar of how innovative solutions can mediate competing interests, promoting cooperation on an international scale.

Keywords: Blockchain Data Management · Blockchain Data Distribution · Paris Agreement · Decarbonization

1 Introduction

To meet the objectives of the Paris Agreement (UNFCCC 2015), all member states of the European Union (EU) jointly committed to reducing greenhouse gas emissions by at least 40% below 1990 levels by 2030. The transport sector is critical to achieving this goal, as it represents more than a quarter of European CO_2 emissions, with road transportation alone responsible for over 70% of transport sector emissions (EEA 2019). In this context, the EU Commission mandates through Regulation (EU) 2019/631 (EU 2019) that starting in 2021 all new light-duty vehicles must be equipped with On-Board Fuel Consumption Meter (OBFCM) devices, which measure and collect individual vehicle data on distance travelled and fuel consumption.

M. Younas et al. (Eds.): Deep-BDB 2023, LNNS 768, pp. 3–15, 2023.
https://doi.org/10.1007/978-3-031-42317-8_1

However, although member states share a common understanding of what needs to be done and thus a common objective, they also have divergent interests. Political organizations such as the EU have to deal with these differences, along with increasing economic, social and ecological trade-offs, as member states compete and collaborate on markets at the same time. Some countries, for example, may wish to protect a strong export-oriented automotive industry, while others that are more affected by global warming at their coastlines may look for stricter reductions in CO_2 emissions, which leads to different actions on global markets. For example, current vehicle emission certification procedures, driven by the economic interests of the automotive industry, underestimate vehicles' emissions by between 30 and 50% (Duarte et al. 2016; Fontaras et al. 2017; Todts 2018). Other examples are the "dieselgate" scandal (European Court of Auditors 2019) and the generally low quality of emissions data in the official European Environment Agency (EEA) database (Kollamthodi et al. 2015).

A supranational organization such as the EU is critical for aligning the actions of nations despite their diverging interests. The CO_2 emission metering system on a European level is an example of this dilemma: countries must cooperate if they are to significantly reduce CO_2 emissions while at the same time, they wish to pursue their own, often competing economic interests. This is also the reason why countries wishing to manage their CO_2 emissions are unwilling to render their data sovereignty and decision power to a centralized system. This situation can be described as coopetition at an international level.

We propose a distributed ledger technology (DLT) based system to enable data sharing that is both transparent and decentralized. While both DLT and blockchain operate on the principle of decentralized consensus to validate transactions, blockchain is a specific type of DLT characterized by sequential chaining of blocks, whereas DLT serves as a broader term encapsulating various forms of distributed databases not necessarily following a block-chained data structure. In the remainder of this paper, we will consistently use the term DLT inclusively address the entire spectrum of technologies enabling distributed data architectures. Shared information systems that monitor and predict CO_2 emissions are essential for smart government decision-making (Tang et al. 2020). DLT allows for an innovative, decentralized, verifiable, and transparent data monitoring system that establishes a "trust-free system" among participants (Beck et al. 2016; Schletz et al. 2020; Treiblmaier 2018). Based on these features, we explore how new coopetition forms at the international level can be enabled. Thus, our research question is *how can a DLT system enable countries to manage and ultimately reduce CO_2 emissions?*

This paper is structured as follows. Section 2 discusses the literature background on DLT systems. Section 3 describes the design science methodology applied, and Sect. 4 will introduce the developed DLT prototype. Section 5 discusses the implications of the prototype evaluation. Section 6 provides a conclusion.

2 Literature Background

In this section, we give an overview of coopetition theory, introduce DLT, and discuss DLT as a foundational technology for coopetition among sovereign nations. We use coopetition theory to analyze the case of a DLT-based project to meter CO2 emissions on European roads.

DLTs provide a decentralized, immutable, and tamper-resilient log, which offers all network participants access to all ledger information (Beck et al. 2018). These features make DLT a potential coopetition enabler if correctly implemented. DLTs achieve disintermediation from trusted third parties through consensus mechanisms and protocol rule enforcement; these mechanisms allow network participants to interact with each other even in the absence of trust (Bano et al. 2017), creating a context in which parties motivated by contrasting economic incentives can safely interact or even cooperate. Network participants (i.e., nodes) need to agree on what constitutes valid transactions. DLT uses cryptography, timestamping, and hashing to record all transactions in a chronological chain that is permanent and extremely difficult to defraud (Narayanan and Clark 2017); over and above that, smart contracts automate the coordination under predetermined transparent rules that are not directly dictated by a single entity, but instead agreed upon by the interacting users. The rapid development of IoT (Internet of Things) technologies, such as the on-board units in this case, facilitate analysis and prediction in government policymaking (Ismagilova et al. 2019). DLT supports the integration and dissemination of this IoT technology data (Dai and Vasarhelyi 2017).

In DLT systems design, there are trade-offs between scalability, security, and decentralization (Yu et al. 2018). Scalability, security, and decentralization also depend on the ownership and accessibility of a DLT system. The ability to submit new transactions and access the stored data in the DLT system is determined by the type of DLT protocol. In permissionless public DLTs, all nodes can validate transactions and maintain the ledger, while in permissioned public or private DLTs, only nodes that have been preregistered and approved can fulfill these tasks (Peters and Panayi 2016). As permissionless and permissioned DLT systems can employ different consensus mechanisms, the type of DLT affects scalability, security, and the degree of decentralization of the system. Which type of DLT system is most suitable for a given task or process can be identified following the decision path developed by Pedersen et al. (2019). Helliar et al. (2020) found that permissioned DLTs generally lag behind permissionless DLTs in terms of diffusion.

3 Design Science Research Methodology

In this research, we follow a design science research (DSR) methodology by constructing and evaluating an IT artefact (March and Smith 1995; Orlikowski et al. 2001), while also building a knowledge base that can guide future artefact design in related areas (Gregor and Hevner 2013; Gregor and Jones 2007). We add to the current stock of DLT design knowledge (which is fairly scant) by developing some first insights (Gregor and Hevner 2013) for applications in decentralized, multi-jurisdictional environments. We follow the four-step guidelines for theory-generating design science research by Beck et al. (2013): (1) creating awareness of the problem and suggesting an approach to solve it; (2) developing the artefact; (3) evaluating the artefact; and (4) abstracting design knowledge. In theory-generating design science research, the artefact should have practical relevance, and its development should be influenced by both the environment (people, organizations and technology) and the knowledge base (foundations and methodologies); therefore, the researcher needs to be well informed when building the artefact (Hevner et al. 2004).

First, we worked with the European Commission's Joint Research Centre (JRC), which initially suggested the problem and outlined the details of our use case. Based on their input, we derived tentative design requirements, which were reiterated in several discussions with experts from the JRC. The JRC subject matter experts were available during the development, evaluation, and theory-generation stages of the research. They helped with open questions regarding the use case and design requirements and provided feedback and insights.

The development and evaluation of a DLT artefact enabled the improvement of design characteristics in response to use case requirements, but also enabled a better understanding of the application's potential and limitations. Further iterative developments took place after the development sprint to further improve the design. Naturalistic evaluation with real vehicles in a real-world driving scenario was not possible within the scope of this research. We implemented a testing environment considering the logic followed to record, store, and interact with the information processed by the envisioned DLT system. Therefore, we chose an evaluation approach with a focus on formative and artificial evaluation methods (Venable et al. 2016).

4 DLT Prototype Construction and Evaluation

This section describes the construction and evaluation of a DLT artefact for the EU transport emission monitoring case study. The aim is to demonstrate how DLT can act as a foundational layer to overcome the challenges posed by coopetition among EU member states.

4.1 Prototype Design Components

The developed prototype fundamentally encompasses two integral components: the DLT network, and OBFCM units affixed within road vehicles, epitomizing a practical application of digital monitoring, reporting, and verification (D-MRV). In this prototype, the OBFCM devices perform the role of capturing source data related to vehicular fuel consumption, thereby providing an accurate and real-time monitoring mechanism. The DLT network, on the other hand, constitutes the reporting element of this structure. Leveraging the inherent attributes of blockchain/DLT technology, it ensures the amplification of data availability, quality, and transparency. Each data input, once recorded onto the DLT network, is practically immutable and readily accessible, thereby enhancing the reliability and credibility of reported information. Moreover, the intrinsic features of DLT, such as decentralization and immutability, impart increased security to the system by reducing the risk of data tampering and fraud. The synergistic interaction between the OBFCM units and the DLT network thus forms an efficient, transparent, and tamper-resistant system for fuel consumption metering and reporting, significantly contributing to the enforcement of carbon emission regulations.

We deem a permissioned public DLT architecture to be the most suitable network configuration for this application. In such a permissioned DLT system, all EU member states and eligible agencies share the ownership of the system by distributing the network's controlling nodes equally amongst themselves, creating accountability and fostering collaboration. These infrastructural network nodes maintain the system's status by ordering and validating data entries and recording them permanently in the DLT system. In a permissioned DLT system, access to data, both for reading and writing, is brokered by the peer nodes. Primarily, peer nodes maintain the ledger of transactions and the state of the network. Each peer node in the network possesses a copy of the ledger, facilitating decentralization and redundancy. This characteristic contributes to the system's resilience and fault-tolerance, ensuring data availability even if some nodes are compromised or offline. In addition, peer nodes in Hyperledger Fabric also perform chaincode (also known as smart contracts) operations. Chaincode, when installed and instantiated on a peer, allows the peer to endorse transactions and interact with the ledger. This functionality is instrumental in processing and validating transactions based on the defined business logic in the chaincode. This means that new information sent by the client nodes (for example, the vehicles' distance travelled or fuel consumption) must be written to comply with network rules, which are granted and enforced by the peer nodes. The same is true for any requests to read data that must go through the peer nodes. Accordingly, a DLT system can be designed to query functions that provide access to specific data levels, such as aggregated data (Manjunath et al. 2018). In this way, sensitive or confidential data will be accessible only to individuals with the required authorization. Other data queries could be used to access information to be reported to the EEA or other relevant environmental or statistical agencies.

For the implementation and evaluation of a DLT-based emission monitoring system, we used Hyperledger Fabric. Hyperledger Fabric is based on three types of network actors: (i) clients, (ii) peers, and (iii) orderers. Each of these actors has a verified identity within the DLT system and is in charge of performing specific tasks. The initial transactions are proposed by client nodes to a subset of peer nodes, according to so-called endorsement policies. Once this subset of peer nodes has validated the transaction, the client nodes submit the information to orderer nodes; these orderer nodes reach consensus on the sequence of transactions, package the information into a single unique new block, and send it to all peer nodes in the system, thereby updating the ledger. Peer nodes hold the transaction log, i.e., the chain of blocks, as well as the smart contracts that automatically execute the application when correctly invoked.

In our case, the DLT system is organized as follows: each of the EU member states and the agencies or institutions representing them (e.g., the Ministry of Transport), as well as European institutions (e.g., the European Commission and the EEA), will own at least one peer node and one orderer node. This means that each country will participate in the computation and validation of transactions from the blockchain infrastructural layer. The system's client nodes will be the vehicles themselves; these are uniquely identifiable as belonging to a specific country through its national vehicle registration system. In this distributed system, each state or designated agency can be certain that each member state is accurately participating in the CO_2 emission monitoring system, and all information is synchronized at the same time to eliminate information asymmetry, providing a consistent and reliable source of data.

The other component of this system is the vehicle's on-board unit that sends transactions containing vehicle and emission information. Current EU on-board unit specifications (EUC 2019) require that units transmit information about the distance travelled and fuel consumption. This data is used to derive the kilometric efficiency and CO_2 emissions for the vehicle (Grant et al. 2008). After the vehicle travels a predefined distance, the on-board units upload their individual verified data directly to the DLT system to prevent potential manipulation by third parties.

Based on cryptographic authentication procedures, the individual vehicle can be identified, and only the specific vehicle can use the designated public key to submit transactions. The permissioned DLT system connects on-board units with the vehicle's specific characteristics (such as manufacturer, model, and fuel type). The DLT system receives transactions, verifies them, and then updates the system's state accordingly. This design provides a secure accountability system for recording vehicle metrics and transparently tracking performances of individual vehicles.

4.2 Prototype Functional Logic Components

In the DLT system, the on-board unit of each vehicle acts as an individual transactive node. For each node, the DLT system contains a specific state entry, as well as all successive transactions of the specific vehicle. This aggregated series of transactions provide a clear view of the vehicle history in terms of kilometers travelled and fuel consumed. In our implementation, transactions can be twofold. The first transaction records are travelled kilometers (KmTx) and are conducted based on distance (for example, each 100 km). The second transaction is the gas station transaction (GsTx) that registers purchased fuel each time a vehicle refuels at a gasoline station (Fig. 1). Each transaction represents a discrete event containing aggregated information about vehicles' metrics since the previous transaction (that is, that the vehicle has travelled 100 km, or that a certain quantity of fuel has been added). By analyzing the specific vehicle data entry, it is possible to extrapolate the total number of travelled kilometers, total fuel consumed, and average fuel efficiency. The DLT system assures the consistency of data and uses smart contracts to enforce the monitoring rules.

Figure 1 illustrates the steps of a fuel purchasing transaction. The procedure for recording fuel consumption starts in parallel with the refueling process. After payment, the data regarding the purchased liters of fuel is recorded by the vehicle's on-board unit. Fuel verification requires a verification not only from the vehicle but also from the gas station, so rather than having the gas station send data to the DLT system separately, the data triangulation between the vehicle and the gas station is established as follows: Vehicle and gas station exchange a signed payload certificate that travels from the vehicle to the gas station (in the form of VSD = Sign(SHA256(liters, time-stamp))), and then back to the vehicle again (in the form of GSSD = Sign(SHA256(VSD))). Finally, the transaction data is uploaded to the DLT system. In addition, the gas station stores the information as proof that the vehicle confirmed that it received a certain amount of fuel at a specific time. If necessary, that information can be triangulated with the uploaded data to detect any manipulation or fraudulent behavior. For such a system to work correctly, gas stations must be able to communicate with vehicles' on-board units. While this technology is not in place yet, gas station providers are already working on such an infrastructure (see (Deutsche Tamoil GmbH 2020)).

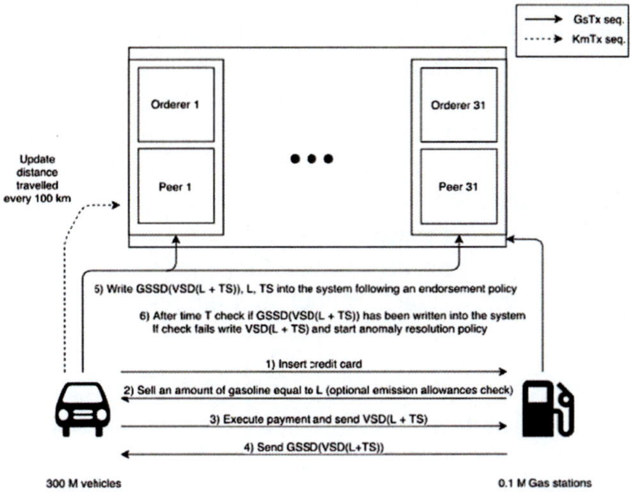

Fig. 1. Architecture and sequence diagram of the emission monitoring system

In summary, the process outlined in Fig. 1 contains a multi-step process for the uploading of OBFCM data to the network of peer nodes, involving client applications and chaincode transactions (Androulaki et al. 2018; Dorri et al. 2017; Hyperledger Fabric 2023):

1. Data Collection: The OBFCM devices installed on vehicles continuously monitor and record data related to fuel consumption.
2. Client Application: A client application, which could be a dedicated software application running on a vehicle's onboard system or a remote server, retrieves this data from the OBFCM devices.
3. Transaction Proposal: The client application creates a transaction proposal, which includes the OBFCM data, and sends this proposal to the endorsing peers on the HF network.
4. Endorsement: The endorsing peers receive the transaction proposal, simulate the transaction using the installed chaincode, and if the transaction is valid, endorse it.
5. Orderer: The endorsed transaction is sent back to the client application, which then forwards it to the orderer node.
6. Block Creation & Distribution: The orderer node packages multiple transactions into a block and then distributes this block to all peer nodes in the network.
7. Validation & Commitment: Each peer node validates the transactions within the block against the network's endorsement policy and, if valid, commits the block to its ledger. Thus, the OBFCM data is now uploaded and stored on the HF network.

4.3 Prototype Evaluation

We evaluated the prototype using vehicle data from the EEA database [EEA 2020]. This database provides detailed information about the manufacturer, model, mass in running order (kg), and the specific CO_2 emissions in g/km for a specific vehicle model. To ensure comparability between the different vehicle models, we used data only from vehicles with gas combustion engines in our evaluation. Based on the specific CO_2 emissions in g/km we calculated the gas consumption in l/100 km.

We generated a simulated vehicle population of N = 500, in which the model is randomly sampled from a set of 10 possibilities. The script simulates the vehicle behavior in terms of distance travelled and fuel consumption, which is represented by transactions, KmTx and GsTx respectively, that are periodically submitted to the testing DLT system. Based on the KmTx and GsTx transactions, CO_2 emissions (g/km) are calculated using the EEA emission factor [Ntziachristos and Samaras 2019]. The data of each individual vehicle is aggregated by vehicle model, and the results of the simulation data are displayed in Table 1.

Table 1. Test vehicle types and aggregated data.

Model	N	Distance (km)	Fuel (l)	Consumption (l/100 km)	Emissions (g/km)
Audi-A4	59	644,650	39,324	6.1	145
Audi-Q2	55	624,350	31,218	5	119
Fiat-500	51	555,450	27,217	4.9	116
Fiat-500L	41	468,850	28,131	6	143
Ford-Fiesta	48	535,500	22,491	4.2	100
Ford-Focus	46	515,600	23,718	4.6	109
Nissan-Micra	55	598,550	25,738	4.3	102
Nissan-Qashqai	54	583,200	31,493	5.4	128
VW-Golf	46	524,300	26,739	5.1	121
VW-Tiguan	45	471,750	30,192	6.4	152

Based on this simulation data, we generated the vehicle emission data. Figure 2 depicts this simulation data to identify vehicle types that are currently complying with the EU fuel consumption efficiency standards (below the orange average line) or that currently do not comply (above the orange average line) and thus will be charged with a penalty.

The prototype automates the integration of data from the on-board units in the DLT system and enables detailed monitoring of individual vehicle emissions, as well as aggregated data by vehicle manufacturer or by country. Countries stay in control of the data and thus can protect information about individual vehicles, while the system will report to all nodes how many vehicles are not in compliance and the size of the excess penalty fee. In this way, CO_2 emissions monitoring is enforced, while sensitive individual data remains protected.

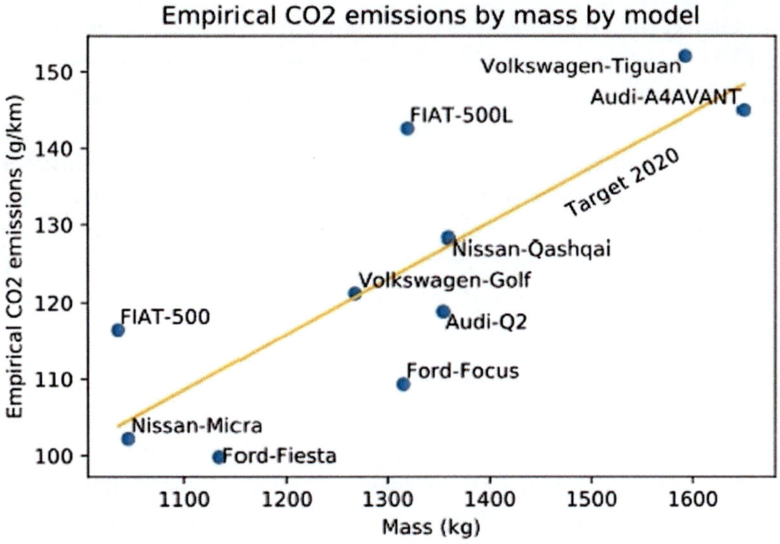

Fig. 2. CO2 emission data compared to vehicle mass

5 Discussion of Empirical Findings

The proposed DLT prototype supports coopetition among participating sovereign member states. While states work collectively on monitoring and reducing CO_2 emissions, and thus enhancing fair cooperation through transparent data sharing, the information pertaining to individual vehicles is kept private, thanks to the permissioned nature of the implemented system. As a result, the monitoring system provides a complete and correct record of the real fuel consumption of each vehicle made by the manufacturer, allowing authorities to enforce the policy-based incentives supporting the joint EU transport sector emission goal. Accordingly, the suggested DLT system enables coopetition by maintaining national data sovereignty and increasing trust, despite ongoing competition. The coordination of policy actions through a shared information base is key to achieving the shared EU objectives.

Our system allows the tracking of individual vehicles and automates data validation through gas station triangulation. The system is a significant improvement compared to the existing fragmented EU emission data management systems. The availability of nearly real-time data for individual vehicles has several practical implications for the entities and actors considered in this paper.

The DLT system provides regulators and policymakers with direct feedback to improve the market mechanism design and provide stronger incentives for reducing CO_2 emissions in a coordinated and cost-effective manner. Currently, it is difficult to assess and plan future policies as data quality is insufficient, and the reduced emissions predicted in certification procedures do not translate into actual emission savings (Fontaras et al. 2017). Poor data quality and the"dieselgate" scandal clearly show that legacy data management processes and systems do not address the challenges of a coopetitive environment.

In the DLT system, uniform data collection and verification methods are automated across all EU member states, accountability is enhanced, and the pressure exerted by European law becomes stronger. Also, the pricing of more harmful vehicles could become more nuanced with an adjusted fee structure, instead of the fixed EUR 95.- per gram of CO_2. Better data quality could allow for the design of market mechanisms to improve the effectiveness of policies and incentivize the introduction of new fuel efficiency technologies. Such granular and transparent action is key to achieving the joint objectives of the EU climate contributions, as governance transparency is vital for ensuring trust and accountability (Pappas et al. 2019).

6 Conclusions

DLT is a harmonizing technology for enabling coopetition at an international level. Despite states' often diverging and heterogeneous interests, the system provides a shared information base, guarantees each party ownership and control over their own data, yet enforces commonly agreed rules across legally independent actors. This "trusted" data layer allows for coordinated action while maintaining national data sovereignty.

In our research, we focused on designing, developing, and evaluating a DLT prototype for emission monitoring on European roads. However, as DLT remains a nascent technology, more empirical testing is required. The developed prototype is at a proof-of-concept level, and we followed an empirical testing approach that in the absence of naturalistic testing relies on expert interviews with the European Commission JRC to assess the robustness of the artefact and its practical use and usefulness.

To empirically evaluate the scalability of this DLT approach, a large-scale network of distributed nodes would be required. Any EU-wide emission monitoring system would need to handle several million vehicles. In times of high transaction loads, scalability limitations might potentially delay the execution of transactions. Our general architecture is platform-agnostic and thus can be applied to any DLT. For our practical illustration of the DLT system, we used Hyperledger Fabric, which offers an "end-to-end throughput of more than 3,500 transactions per second in certain popular deployment configurations, with sub-second latency, scaling well to over 100 peers" (Androulaki et al. 2018, p. 1). If we assume a total of 300 million vehicles, with an average annual mileage per vehicle of 15,000 km and an average reporting interval per vehicle of 1,000 km, this would result in approximately 150 transactions per second. This number will fluctuate, being significantly higher during rush hours and lower during times of low traffic. Hyperledger Fabric's 3,500-transaction-per-second capacity limit is adequate to handle this level of transactions. Also, it is fair to assume that the technology will mature even further, and new DLT technologies will present further improvements (Glaser 2017).

All tests were based on Hyperledger Fabric testing tools. Thus, while we have been able to test and assure functional integrity, an evaluation of the system's latency time due to the difference in the size of the data uploads, or congestion caused by time-of-day fluctuation of data submissions, is yet to be done. Another potential issue is that in the current implementation, the available ordering-service consensus algorithms are only CFT (Crash Fault Tolerant) and not BFT (Byzantine Fault Tolerant). While this could create a problem in other implementations, in our case, all ordering nodes of the permissioned public emission monitoring system are controlled by known entities such as the EU member states. The likelihood that a BFT attack would occur in our implementation is very low, however, in general, a BFT algorithm would be a better choice for applications that do not require trust.

References

Androulaki, E., et al.: Hyperledger fabric: a distributed operating system for permissioned blockchains (2018). ArXiv:1801.10228. https://doi.org/10.1145/3190508.3190538

Bano, S., Sonnino, A., Al-bassam, M., Azouvi, S., Mccorry, P.: SoK: Consensus in the Age of Blockchains (2017). ArXiv: 1711.03936. https://arxiv.org/pdf/1711.03936.pdf

Beck, R., Müller-Bloch, C., King, J.L.: Governance in the blockchain economy: a framework and research agenda. J. Assoc. Inf. Syst. **19**(10), 1020–1034 (2018). https://doi.org/10.17705/1jais.00518

Beck, R., Stenum Czepluch, J., Lollike, N., Malone, S.: Blockchain - the gateway to trust-free cryptographic transactions. In: 24th European Conference on Information Systems, ECIS 2016, pp. 1–14 (2016)

Beck, R., Weber, S., Gregory, R.W.: Theory-generating design science research. Inf. Syst. Front. **15**(4), 637–651 (2013). https://doi.org/10.1007/s10796-012-9342-4

Dai, J., Vasarhelyi, M.A.: Toward blockchain-based accounting and assurance. J. Inf. Syst. **31**(3), 5–21 (2017). https://doi.org/10.2308/isys-51804

Deutsche Tamoil GmbH. "Pay at pump"-Smart fueling using Connected Fueling HEM and PACE cooperate in mobile payment at the fuel pump (2020). https://www.pace.car/press/press_releases/200602_-_HEM_and_PACE_cooperate_in_mobile_payment_at_the_fuel_pump.pdf

Dorri, A., Kanhere, S.S., Jurdak, R., Gauravaram, P.: Blockchain for IoT security and privacy: the case study of a smart home. In: 2017 IEEE International Conference on Pervasive Computing and Communications Workshops (PerCom Workshops), 618–23. IEEE (2017). https://doi.org/10.1109/PERCOMW.2017.7917634

Duarte, G.O., Gonçalves, G.A., Farias, T.L.: Analysis of fuel consumption and pollutant emissions of regulated and alternative driving cycles based on real-world measurements. Transp. Res. Part D: Transp. Environ. **44**, 43–54 (2016). https://doi.org/10.1016/j.trd.2016.02.009

EEA. Greenhouse gas emissions from transport sector in Europe. European Environment Agency, 2 (2019). https://www.eea.europa.eu/data-and-maps/indicators/transport-emissions-of-greenhouse-gases/transport-emissions-of-greenhouse-gases-12

EUC. REGULATION (EU) 2019/631 OF THE EUROPEAN PARLIAMENT AND OF THE COUNCIL setting CO2 emission performance standards for new passenger cars and for new light commercial vehicles, and repealing Regulations (EC) No 443/2009 and (EU) No 510/2011. Official Journal of the European Union (2019). https://eur-lex.europa.eu/legal-content/EN/TXT/PDF/?uri=CELEX:32019R0631&from=EN

European Court of Auditors. The EU's response to the "dieselgate" scandal. Briefing Paper (2019). https://www.eca.europa.eu/lists/ecadocuments/brp_vehicle_emissions/brp_vehicle_emissions_en.pdf

Fontaras, G., Zacharof, N.-G., Ciuffo, B.: Fuel consumption and CO2 emissions from passenger cars in Europe – laboratory versus real-world emissions. Prog. Energy Combust. Sci. **60**, 97–131 (2017). https://doi.org/10.1016/j.pecs.2016.12.004

Glaser, F.: Pervasive decentralisation of digital infrastructures: a framework for blockchain enabled system and use case analysis. In: HICSS 2017 Proceedings, pp. 1543–1552 (2017). https://doi.org/10.1145/1235

Grant, M.D., Choate, A., Pederson, L.: Assessment of Greenhouse Gas Analysis Techniques for Transportation Projects. The National Academics of Sciences (2008). https://trid.trb.org/view/848709

Gregor, S., Hevner, A.R.: Positioning and presenting design science research for maximum impact. MIS Q. Manag. Inf. Syst. **37**(2), 337–355 (2013). https://doi.org/10.25300/MISQ/2013/37.2.01

Jones, David, Gregor, Shirley: The anatomy of a design theory. J. Assoc. Inf. Syst. **8**(5), 312–335 (2007). https://doi.org/10.17705/1jais.00129

Helliar, C.V., Crawford, L., Rocca, L., Teodori, C., Veneziani, M.: Permissionless and permissioned blockchain diffusion. Int. J. Inf. Manag. **54**, 102136 (2020). https://doi.org/10.1016/j.ijinfomgt.2020.102136

Bichler, Martin: Design science in information systems research. MIS Q. **48**(2), 133–135 (2006). https://doi.org/10.1007/s11576-006-0028-8

Hyperledger Fabric. A Blockchain Platform for the Enterprise. Hyperledger Fabric (2023). https://hyperledger-fabric.readthedocs.io/en/release-2.5/whatis.html. Accessed 13 June 2023

Ismagilova, E., Hughes, L., Dwivedi, Y.K., Raman, K.R.: Smart cities: advances in research—an information systems perspective. Int. J. Inf. Manag. **47**(January), 88–100 (2019). https://doi.org/10.1016/j.ijinfomgt.2019.01.004

Kollamthodi, S., Kay, D., Skinner, I., Dun, C., Hausberger, S.: The potential for mass reduction of passenger cars and light commercial vehicles in relation to future CO2 regulatory requirements: Appendices. Ricardo-AEA. Report for the European Commission – DG Climate Action, Ref. CLIMA (2015). https://ec.europa.eu/clima/sites/clima/files/transport/vehicles/docs/ldv_downweighting_co2_appendices_en.pdf

Manjunath, P., Soman, R., Shah, D.P.G.:. Iot and block chain driven intelligent transportation system. In: Proceedings of the 2nd International Conference on Green Computing and Internet of Things, ICGCIoT 2018, pp. 290–293 (2018). https://doi.org/10.1109/ICGCIoT.2018.8753007

March, S.T., Smith, G.F.: Design and natural science research on information technology. Decis. Support Syst. **15**(4), 251–266 (1995). https://doi.org/10.1016/0167-9236(94)00041-2

Narayanan, A., Clark, J.: The concept of cryptocurrencies is built from forgotten ideas in research literature. Commun. ACM **15**(4), 1–30 (2017). https://doi.org/10.1145/3132259

Orlikowski, W.J., Lacono, S.C.: Research commentary: desperately seeking the "IT" in IT research - a call to theorizing the IT artifact. Inf. Syst. Res. **12**(2), 121–134 (2001)

Pappas, I.O., Mikalef, P., Dwivedi, Y.K., Jaccheri, L., Krogstie, J., Mäntymäki, M. (eds.): Digital Transformation for a Sustainable Society in the 21st Century: 18th IFIP WG 6.11 Conference on e-Business, e-Services, and e-Society, I3E 2019, Trondheim, Norway, September 18–20, 2019, Proceedings. Springer, Cham (2019). https://doi.org/10.1007/978-3-030-29374-1

Pedersen, A.B., Risius, M., Beck, R.: Blockchain decision path: "when to use blockchain?" – "which blockchain do you mean?". MIS Q. Execut. **18**(2), 24 (2019). https://pure.itu.dk/ws/files/83594249/MISQe_BC_in_the_Maritime_Shipping_Industry_Revision.pdf

Peters, G.W., Panayi, E.: Understanding modern banking ledgers through blockchain technologies: future of transaction processing and smart contracts on the internet of money. New Econ. Wind., 239–278 (2016). https://doi.org/10.1007/978-3-319-42448-4_13

Schletz, M., Franke, L., Salomo, S.: Blockchain application for the paris agreement carbon market mechanism – a decision framework and architecture. Sustainability **12**(5069), 1–17 (2020). https://doi.org/10.3390/su12125069

Tang, M., Wang, S., Dai, C., Liu, Y.: Exploring CO2 mitigation pathway of local industries using a regional-based system dynamics model. Int. J. Inf. Manag. **52**, 102079 (2020). https://doi.org/10.1016/j.ijinfomgt.2020.102079

Todts, W.: CO2 Emissions From Cars: the facts. Brussels (2018). https://www.transportenvironment.org/sites/te/files/publications/2018_04_CO2_emissions_cars_The_facts_report_final_0_0.pdf

Treiblmaier, H.: The impact of the blockchain on the supply chain: a theory-based research framework and a call for action. Supply Chain Manag. Int. J. **23**(6), 545–559 (2018). https://doi.org/10.1108/SCM-01-2018-0029

UNFCCC. Paris Agreement, United Nations Framework Convention on Climate Change. 21st Conference of the Parties. Paris (2015). https://www.undocs.org/FCCC/CP/2015/L.9

Venable, J., Pries-Heje, J., Baskerville, R.: FEDS: a framework for evaluation in design science research. Eur. J. Inf. Syst. **25**(1), 77–89 (2016). https://doi.org/10.1057/ejis.2014.36

Yu, G., Wang, X., Zha, X., Zhang, J.A., Liu, R.P.: An optimized round-robin scheduling of speakers for peers-to-peers-based byzantine faulty tolerance. In 2018 IEEE Globecom Workshops (GC Wkshps), pp. 1–6. IEEE (2018). https://doi.org/10.1109/GLOCOMW.2018.8644251

Moving Towards Blockchain-Based Methods for Revitalizing Healthcare Domain

Rihab Benaich(✉), Saida El Mendili, and Youssef Gahi

National School of Applied Sciences, Ibn Tofail University, Kenitra, Morocco
{rihab.benaich,elmendili.saida,gahi.youssef}@uit.ac.ma

Abstract. The healthcare industry is widely regarded as one of the most critical domains in contemporary society. It plays a pivotal role in promoting the overall physical, mental, and social well-being of individuals worldwide, and efficient healthcare delivery can significantly contribute to a government's economic development and modernization. However, ensuring the security of vast volumes of healthcare data remains a primary concern. Many traditional solutions are inadequate for addressing this particular issue. Therefore, researchers and practitioners have developed novel methods that rely on cutting-edge technology to achieve a high level of security. One such technology that has gained significant attention is blockchain, owing to its robust properties for instance decentralization, immutability, and transparency. Blockchain has demonstrated promising results in various domains, from the financial and innovative sectors to the healthcare industry. In this paper, we first outline the challenges associated with security and data management in healthcare. Subsequently, we highlight the key components and consensus mechanisms of blockchain technology. We then present an overview of blockchain applications in the healthcare field by reviewing relevant use cases, such as improving patient monitoring, controlling pharmaceutical supply chains, and ensuring the security of electronic medical records. Ultimately, we identify the constraints and challenges of existing research and provide recommendations for future works.

Keywords: Blockchain · Healthcare · Security · Decentralized technology · Big data · Data management

1 Introduction

Currently, healthcare is one of the most eminent and leading domains that affect the whole global population and is nearly related to the growth of any country. It also significantly impacts how a country is seen in terms of economic stability. Nevertheless, as healthcare spending rises and business evolves, security and accessibility remain significant issues. Unfortunately, numerous healthcare security flaws might threaten patient data. Without proper management, electronic health data and other vital information can easily slip into the hands of malevolent individuals. Furthermore, the duality of shared healthcare information is that it both makes patients safer and puts them in danger. The more

M. Younas et al. (Eds.): Deep-BDB 2023, LNNS 768, pp. 16–29, 2023.
https://doi.org/10.1007/978-3-031-42317-8_2

comprehensive the network, the more beneficial it is in providing high-quality medical treatment; nonetheless, its data becomes more appealing to fraudsters. According to IBM research, the average cost of a significant data breach in 2020 will be 4.24 *million, costing the healthcare industry* 499 million [1].

Moreover, the COVID-19 pandemic has made safeguarding hospitals and healthcare institutions more complicated by introducing additional worries about disease transmission to the extensive list of other dangers that healthcare security personnel confront. For this reason, many solutions and procedures were tremendously required to face security and data management issues [2]. These solutions have positively impacted the sector by introducing main methods and models. However, numerous of them aren't entirely sufficient, especially concerning the security challenge. Currently, many traditional solutions still rely on centralized models, increasing unauthorized access of individuals and malicious activities and expelling sensitive patient data. Therefore, there is an immense demand for high-quality healthcare facilities backed by innovative technology to overcome the security challenge [3]. One of these relevant technologies is blockchain. Satoshi Nakamoto invented it in 2008 [4]. Since then, blockchain technology has advanced over the past decade, revolutionizing how corporate operations are documented in the digital environment.

Blockchain technology is described as a shared, decentralized, immutable ledger that helps secure the transaction process across multiple devices without the involvement of a central authority. The decentralized technology has developed over the following three generations:

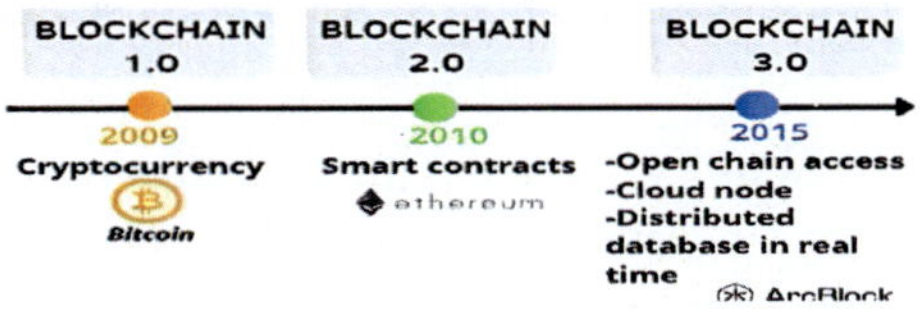

Fig. 1. Blockchain generations.

The first generation was Blockchain 1.0, which was built on Bitcoin. This first blockchain implementation focused on cryptocurrency applications. The second version, Blockchain [5], has evolved with the notion of smart contracts, which are designed, performed, and stored in the distributed ledger. For instance, in Blockchain 3.0 [6], this third generation is primarily concerned with non-financial applications, for example, energy and healthcare. With the rising interest in blockchain and its adoption in various industries, including healthcare, supply chain, banking, and even digital entertainment, healthcare has emerged as a prominent area where different blockchain usage scenarios have been established. In this contribution, we highlight the correlation between blockchain and healthcare and how this intelligent technology has impacted the healthcare sector by first presenting the core components and types of blockchain, then by reviewing the new and relevant studies done in this context, and lastly by discussing the significant benefits and challenges brought by the decentralized technology.

The ongoing paper is presented as follows: in the second and third sections, firstly, we underline our research methodology, then we highlight our research problem. Then, in the fourth section, we focus on overviewing blockchain technology, first by highlighting the fundamental concepts of the technology; Fig 1. Blockchain generations then, we emphasize the central consensus mechanisms and the different types of blockchains. Afterward, in Sect. 5 and Sect. 6, we underline the importance of blockchain integration in healthcare services by presenting the novel and pertinent research is done in this context. Then, in Sect. 7, we discuss the outstanding benefits brought by blockchain to healthcare and the significant challenges. Lastly, we conclude the paper, and we highlight our future work. The following section outlines the blockchain generalities, for instance, blockchain representation, key concepts, and the types of blockchain technology.

2 Blockchain Technology Fundamentals

The digital ledger, i.e., blockchain, is described as a distributed ledger that records all P2P network transactions and where participants can confirm transactions without needing a centralized authority or an intermediary. A blockchain is represented as a chain (series) of blocks. This chain is started by the initial block, the genesis block. In- depth, the blocks Fig. 2. That compose the chain are cryptographically linked. This implies that each block includes the hash value of its previous block. This method makes decentralized technology one of the most secure and robust against vulnerabilities.

2.1 Key Concepts

Since the blocks are the backbone of the blockchain, it's necessary to highlight the significant elements composing these blocks. Among these elements, these are the key ones:

- **A magic number:** is a number that contains specified values that identify a block as being part of a given cryptocurrency's chain. Also, it certifies the block's beginning and confirms that the data originates from the production network [7].
- **Block size:** it stands for the limit size of the block so that it can only hold a certain amount of data.
- **Block header:** this section includes specific information related to the league.
- **A transaction counter:** is a number that reflects the number of records contained in a block.
- **Transactions:** A record of all trades is included inside a partnership.

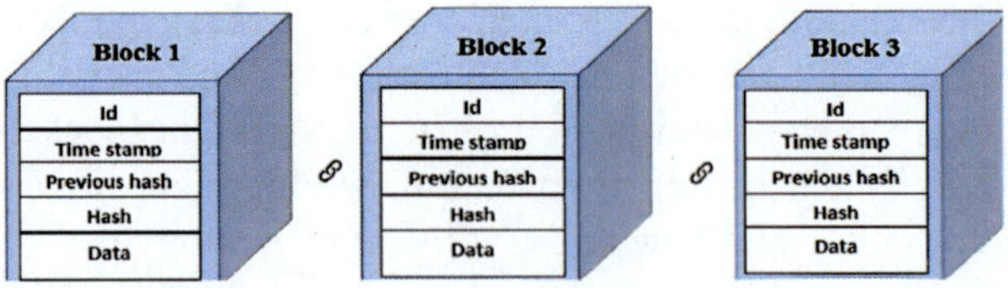

Fig. 2. Blockchain representation.

Furthermore, the blocks contain the most significant transactions because they carry a lot of information. It includes the following sub-elements:

- **Version:** The cryptocurrency version is currently in use.
- **Preceding block hash:** An encrypted hash of the header of the preceding block.
- **Hash Merkle root:** A hash of the transactions in the current block's Merkle tree.
- **Time:** A timestamp for inserting the block onto the blockchain.
- **Bits:** The target hash's difficulty level indicates how tough it is to solve the nonce.
- **Nonce:** is an encrypted number that a miner must decode to validate and close the block.

2.2 Blockchain Taxonomy

Blockchains can be divided into four classes (permissionless blockchain, permissioned blockchain, hybrid blockchain, and consortium blockchain) [8] (Fig. 3):

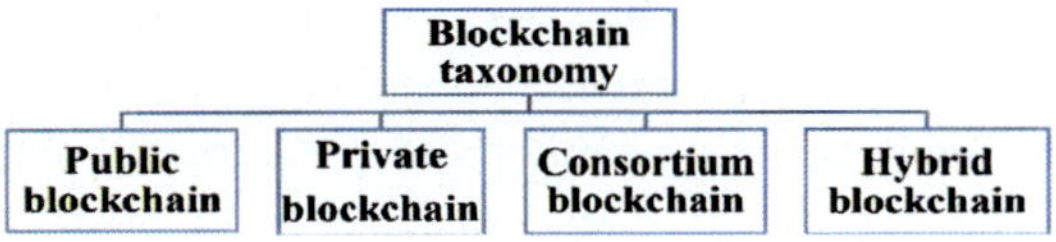

Fig. 3. Blockchain taxonomy.

- **Public or permission-less blockchains:** in this type of blockchain, participants don't require permission to access the network. In a public blockchain, members can participate in the consensus process, read and transmit transactions, and maintain the public blockchain. Moreover, all members may have the right to publish, access, and validate new blocks, allowing them to keep a copy of the whole blockchain. Furthermore, the recent transactions added by members are represented as blocks. These blocks are verified through consensus mechanisms. Among this verification process is a puzzle-solving method used in bitcoin and a staking cryptocurrency method used in Ethereum. The use of consensus mechanisms of blockchain guarantees security.
- **Private or permissioned blockchains:** this type of blockchain is more secure than public blockchains since only trusted participants can join the network through invitation. The private blockchain is dedicated to a unique organization.
- **Consortium blockchain:** this blockchain type is similar to private but with some varieties. Private and consortium blockchains are faster, more protected, and more secure than private blockchains. The Consortium blockchain is dedicated to multiple organizations. It is restricted only to invited participants.
- **Hybrid blockchain:** this type is a union of both private and public blockchains since harmony is used in this specific blockchain. A Hybrid blockchain can benefit from the accessibility of the general ledger and the control level of private.

Undoubtedly, these blockchains have considerable differences in security level, speed, and usability. However, all of them share the main characteristics of blockchain: decentralization, immutability, security, traceability, autonomy, anonymity, integrity, and transparency. These salient features of blockchain have aroused the attention of many researchers to compete and use this technology in different fields. One of them is healthcare. The following section emphasizes the use cases of decentralized technology in healthcare.

3 Blockchain Technology in Service of Healthcare

Since blockchain technology is well-known for its intriguing attributes, especially regarding security concerns, significant industries and fields have adopted it. At first, blockchain gained enormous attention in banking and managing financial transactions. Nevertheless, lately, many sectors, such as healthcare, have oriented their research on blockchain-based solutions. In this section, we first underline the relevant challenges faced in healthcare, and then we highlight the requirement for blockchain technology to substitute traditional methods. Healthcare is one of the most critical and sensitive domains where health information about every patient is essential and must be managed carefully. Medical records are currently maintained in centralized databases managed by individual users, businesses, or large groups of organizations. This centralized management aspect of healthcare data makes it prone to attacks and increases the risk of the malicious user entry. Another challenge faced in healthcare is health information exchange, marked as one of the most time consuming and monotonous activities contributing to high medical costs.

Moreover, the downsides of typical electronic health records include the lack of assurance of the record's data, integrity, or availability and that a third party must administer the database. It is also difficult to connect numerous hospitals to exchange this information and connect them smoothly for the best purpose [9]. Furthermore, data management is also one of the considerable challenges in the healthcare sector. Patients' data must be managed and maintained accurately and meticulously [10]. Hence, employing compelling technologies such as blockchain to overcome the challenges of security and data management is seen as a priority when related to the primary care field. Data in a blockchain is held across networks rather than a central database, boosting stability and demonstrating its vulnerability to hacking [11].

Therefore, the confidentiality of patients' records will be ensured by preventing the alteration of medical history. Because blockchain technology allows patients to make available their medical data and access legislation, it facilitates the move to interoperability driven by patients. Hence, patients will obtain more control over their personal information while improving privacy. Besides the security of health records, health and pharmaceutical industries will use blockchain technology to eliminate fraudulent medications, allowing for the traceability of all used products [12].

4 Related Works

4.1 Research Methodology

This research study emphasized the vital significance of exceptional Blockchain technology in addressing healthcare areas like security and data management [13, 14]. We adopted a systematic review technique to select and summarize the relevant research contributions that address the following questions for this work. We performed our review following Tranfield [15] guidelines to produce a review strategy for the methodologies employed, the primary research questions to be specified, and the reports to be retrieved. This study was carried out in the three following stages:

- First and foremost, we studied and discovered several primary documents relevant to our issue.
- We then filtered the primarily chosen publications.
- Finally, the screened papers were confirmed for data synthesis and analysis.

The significant studies were chosen based on the following criteria, for instance: evaluating publications, the scope of the investigation, and subject and title verification, as well as reading the abstract and findings. This research strategy concentrates on articles published in prestigious academic publications such as Springer, IEEE Xplore, Wiley online library, Google Scholar, ACM digital library, Web of Science, etc. In our study strategy, we explored many articles and publications released in recent years. Three keyword combinations were suitable for the proper database search: blockchain and healthcare, medical health; medical management; or health management. Through source research, numerous articles were discovered. This initial bundle was refined using a relevant and thorough investigation of cross-referenced and redundant keyword combinations. Afterward, the filtered papers were thoroughly examined and evaluated to verify that they met the criteria. The upcoming section highlights our research problem first by presenting the challenges encountered in the healthcare field. Afterward, we will provide an overview of the fundamental objectives we aim to tackle shortly.

4.2 Our Research Foresight Regarding Healthcare Challenges

Since healthcare is one of the most sensitive domains, patients' data confidentiality and privacy are considered top level concerns. For instance, finding an accurate solution based on new technologies that match healthcare requirements, for example, blockchain, is a foremost necessity. The traditional methods and solutions used in this context are no longer adequate, especially when the data growth rate is massive and data breaches are spreading exponentially. Therefore, we aim to provide a blockchain based solution that gathers the sighting properties of blockchain to achieve a secure ecosystem for patients, doctors, and primary care providers while preserving the accurate management of large amounts of data [16]. We also seek to supply solutions for multiple use cases in the healthcare sector, for instance, enhancing patient monitoring, controlling the pharmaceutical supply chain, and ensuring the security of electronic medical records.

4.3 Blockchain Adoption in Healthcare Domain

After highlighting the significant challenges encountered in the healthcare field and the capacity of blockchain to conquer security and data management issues, this section highlights the relevant studies and research that have tackled the utilization of blockchain technology in healthcare. Blockchain has been successfully used in many scenarios in the healthcare field, such as ensuring security and improving data management. In this context, the authors in [17] suggest a solution for patient-controlled security (PHR administration) that is not reliant on a central authority. Their method operates on public and private blockchains and stores data via the IPFS protocol. An intelligent controller contract registers patients, physicians, hospitals, and other healthcare-related entities on the blockchain as nodes. The use of smart contracts in their method enables the request and authorization to access records, which are then recorded as transactions on the ledger. Smart contracts are used for access control and privacy preservation, and the InterPlanetary Name System (IPNS) is used to create a unique identifier for each patient.

A mobile application is also included in the solution, allowing patients to manage their medical records and grant access to healthcare providers. The authors' approach provides a safe and transparent platform for patient data management, with patients always in control of their data. Likewise, the authors in [11] describe a solution based on most blockchain for accelerating care control in terms of administrative procedures and health results. A blockchain based network can swiftly link many participants, such as physicians, scientists, lab workers, and public health authorities. This solution uses smart contracts to guarantee data and treatment process reliability between patients and healthcare experts. Also, the authors of the paper examined the current status of healthcare and identified various concerns, such as data privacy, data security, interoperability, and patient-centric treatment. The researchers looked at how blockchain technology can be utilized to address these issues and enhance healthcare results in terms of medical record management, drug traceability, supply chain, clinical trials, payment and insurance management.

Furthermore, the authors in [18] presented a solution that enables safe medical data interchange across local healthcare institutions, incorporating various national and international entities and correlating critical medical events to develop, manage, and control epidemics. The suggested system is built on blockchain technology, allowing for the customizable design to increase medical data sharing among different health organizations while meeting the quality of service (QoS) expected in health. The proposed system consists of four major components which are: a private blockchain- based data storage system that allows for secure and tamper-proof storage of health records. Also, a privacy-preserving data sharing technique that enables privacy- preserving data sharing between healthcare providers and patients by combining cryptographic techniques such as homomorphic encryption and secure multi-party computation. Another primordial constituent of the system is an access control and data governance solution, implements access control policies and data governance rules using smart contracts. This ensures that only authorized individuals have access to and modify healthcare data, and that data is used in accordance with applicable laws and regulations. Also, patient-centric data management responsible for giving patients control over their data.

In addition to the previous researches, the authors of the paper [19] have proposed a novel approach to addressing data security and privacy challenges in the healthcare industry. The paper presents the Federated Blockchain System (FBS), which uses the unique features of blockchain technology to enable secure and efficient patient data sharing across different healthcare organizations. The system is built on a federated architecture, which means that each organization retains control over its own data while also participating in a larger network for data sharing and collaboration. Also, the paper highlighted how this federated blockchain system advanced encryption mechanisms such as homomorphic encryption and attribute-based encryption. These methods enable secure data sharing while protecting patient confidentiality.

In parallel, many companies have dedicated their solutions to healthcare challenges based on blockchain technology. Patientory [20] is an example of a digital health company that has used blockchain as a pillar solution to handle primary-care problems. The Patientory company has introduced one of the disruptive projects, a decentralized application (Dapp). Patientory's Dapp is an open and secure system that collects transaction records on linked blocks and saves them on a distributed and encoded database that serves as a ledger.

Because the records are distributed throughout a replicated database network, all databases are in line. Blockchain provides great security benefits. On the other hand, users can only access the blocks for which they have authorization. The Patientory's Dapp provides patients access to their health data securely and rapidly. All the studies mentioned above have certainly impacted the healthcare sector, whether by enhancing health data security or facilitating data management. The following table highlights the considerable research and studies that have tackled the leverage of using blockchain in the healthcare field by classifying them into the following major use cases: enhancing patient monitoring, controlling pharmaceutical supply chains, ensuring the security of electronic medical records, and managing health insurance claims (Table 1).

Table 1. The main applications of blockchain integration in healthcare

Use case	Paper	Contribution	Framework/ Model	Advantages	Drawbacks
Remote patient monitoring	[21]	Enhancing disease diagnosis using blockchain and ML	-Permissioned Blockchain	-Security and privacy of patients -Reliability	-Latency -Scalability -Interoperability
	[22]	Increasing data availability between hospitals and doctors by implementing an access policy control algorithm	-Hyperledger Fabric	-Enhancement of healthcare services -Efficiency and security –Data accessibility	-Scalability
Pharmaceutical supply Chain	[23]	It proposes a solution that ensures a clear flow of drugs between entities in the distribution chain.	-Zero-Knowledge proof protocol -Markov model		

(*continued*)

Table 1. *(continued)*

Use case	Paper	Contribution	Framework/ Model	Advantages	Drawbacks
	-Security -Consistency -Availability	-Scalability -Nodes are identified by their IP address instead of anonymized public address			
	[24]	Creation of a blockchain-based control system for medication return management	-Hyperledger Fabric	-Provides Accurate identification of medical products' legitimacy	The system cannot track counterfeit pharmaceuticals provided through ways other than approved channels
Electronic medical records	[25]	Implementation of patient-centric blockchain intelligent contracts that meet patients' and healthcare professionals' needs.	-The system is based on smart contractsand cryptographic hash functions	-Integrity -Security -Interoperability -Transparency	-Latency -Scalability
	[26]	Consortium blockchain technology can contribute to developing new EHR	-Consortium Blockchain	-Privacy preservation -Security	-Latency -Single point of failure
Health insurance claims	[27]	Presentation of the blockchain-based solution to build a method for processing health insurance claims that avoid fraud and risk	-Smart contracts -Proof of work algorithm	-Secure network free from tampering	Network latency

This paper investigates the existing literature thoroughly and addresses the limitations and obstacles encountered in prior research. On the basis of these insights, we present a compelling strategy for addressing the challenges associated with health records using blockchain technology. Furthermore, in this paper we focus on the critical aspect of security and we propose the incorporation of advanced encryption mechanisms to strengthen the protection and privacy of electronic health records. Moreover, the purpose of this paper is to provide valuable insights and potential solutions for advancing the secure use of blockchain technology in the healthcare industry.

5 Discussion

Based on the various studies and research mentioned in this paper, we highlighted how blockchain technology plays a pivotal role in securing and managing patients' data effectively and how it could be a noteworthy alternative to traditional methods considering that these outdated methods are incapable of covering all healthcare field requirements, mostly towards security concerns. Breach incidents are widespread in the healthcare industry. Various circumstances can trigger these, such as credential-stealing malware, an insider who intentionally or unintentionally releases patient data, or misplaced laptops or other devices. Health and human services have exposed more than 15 million health records [28]. Moreover, according to [28], more than 78 million individuals were affected by the Anthem data breach.

Therefore, blockchain technology is considered a compatible solution for healthcare demands. As reported by Statista, around 55% [29] of healthcare applications will use blockchain for commercial deployment by 2025 [30]. Due to blockchain's robust characteristics mentioned earlier, namely immutability, security, decentralization, transparency, consensus, faster settlement, and distributed ledgers. The decentralized technology offers a range of benefits to the healthcare field by introducing smart contracts, consensus algorithms, and cryptographic hash functions appended with blockchain's features. This significant technology has the potential to transform the healthcare industry, particularly in the management of patient data. One of the key benefits of this technology is that it allows patients to have more control over their medical data, including the ability to control who has access to it and even monetize it. Patients now have the ability to take control of their medical records, which represents a significant shift in power dynamics.

Furthermore, because researchers have access to a larger pool of data, the use of blockchain can facilitate more efficient and effective research. Another potential benefit is the simplification of medical data sharing among different healthcare providers, which could lead to better care coordination and patient outcomes. Blockchain technology reduces the administrative burden on healthcare providers. Because of this technology, also the security level is remarkably enhanced, and health information is accurately stored and managed. However, many challenges remain, such as interoperability and scalability difficulties, which are the primary focus of current and future research in the context of blockchain's adoption in healthcare services. Because several developed apps are still unable to connect, the interoperability challenge underlines the absence of standards for developing blockchain-based healthcare apps.

Furthermore, scalability results are a big concern in blockchain-based healthcare systems [31], particularly given the volume of medical data. Because of the massive amount of data produced in healthcare, managing this vast data based on the on-chain method included in the blockchain is not practical since healthcare institutions require large and efficient data storage[16]privacy to avoid performance degradation. Finally, another significant challenge related to blockchain's integration into healthcare ecosystems is latency. This issue is caused by the pace of off-chain data load and transaction processing. Blockchain-based health care systems have the potential to improve patient care, but current research has both technical limitations and strengths. While the immutability and transparency features of blockchain technology improve data integrity, there are issues with interoperability with existing healthcare systems, scalability, and performance when managing large amounts of data on-chain. Furthermore, high gas costs in permissionless blockchains limit the ability to store large amounts of data on-chain.

Therefore, the scope and depth of current studies on blockchain implementation for healthcare systems vary, with some focusing on technical aspects and others on practical implementation and adoption. Despite the promising advances made in blockchain-based EHR research, there is a notable lack of standardization in the methodologies used, which makes comparing and replicating results difficult. Additionally, few studies have provided comprehensive analysis of the broader healthcare ecosystem and the various stakeholders involved in blockchain implementation. These challenges are expected to be addressed as the technology matures and research continues, making blockchain an indispensable component of the healthcare ecosystem. In Fig. 4., we highlight a swot analysis (the strengths, weaknesses, opportunities, and threats) of blockchain adoption in healthcare.

	Strengths	**Weaknesses**
Internal factors	- Data transparency - Security , real-time data storage - Tamper proof, autonomous network	-Scalability -Technical immaturity - Complexity
	Opportunities	**Threats**
External factors	-Cost-efficient -Lower fraud risks -Increased trust of patients -Improved treatment and care	-Interoperability challenges -Lack of knowledge -Hesitation of adoption

Fig. 4. A SWOT analysis of blockchains' integration in healthcare.

6 Our Forthcoming Proposition

We are committed to addressing the challenge of patient record security in our upcoming proposed solution-by leveraging cutting-edge techniques such as blockchain technology, smart contracts, and advanced encryption mechanisms. We recognize the need for a more secure and robust solution to protect patient data in light of the increasing frequency of data breaches and cyberattacks on healthcare systems. Blockchain technology enables the storage and management of patient records in a decentralized and tamper-proof environment. We can use this technology to ensure that all patient data is recorded and secured on a distributed network of nodes, reducing the risk of a single point of failure. Furthermore, the use of smart contracts enables the automation of certain data management tasks, data management responsibilities include record-keeping, data access, and consent management. We will create a set of rules and conditions for data sharing and access using Solidity, a programming language used to write smart contracts on the Ethereum blockchain. This will simplify the process of managing patient records and eliminate the need for middleman.

We are also incorporating advanced encryption mechanisms into the system to increase the level of security even further. This will include the use of cutting-edge encryption algorithms and secure key management protocols to safeguard patient data and ensure its confidentiality at all times. We can provide remarkable levels of protection to patient records by implementing a multi-layered approach to data security. To provide a secure environment for sharing and storing patient data, our proposed solution will rely primarily on a combination of algorithms, including the advanced encryption standard and the zero-knowledge proof protocol. Overall, our upcoming solution will help to improve patient data security and privacy while also facilitating better healthcare outcomes through improved access and data sharing.

7 Conclusion

The massive amount of data requires compelling strategies and methods to be handled accurately. In this context, blockchain, an emerging technology, has shown the ability to provide novel and brilliant solutions to different sectors, including healthcare. Due to its superior properties, blockchain technology can overcome challenges associated with patient security and privacy. It can also be adopted as a great alternative to traditional data storage methods, generally making a qualitative leap in the healthcare field. Besides security and data management aspects, blockchain can effectively minimize costs related to solutions based on central authorities by integrating smart contracts. Despite the relevant advancements observed in blockchain integration in healthcare services, scalability and interoperability are two significant challenges that hamper adopting decentralized healthcare technology. Another challenge encountered and equally crucial as scalability and interoperability is latency. Therefore, more study is needed to adequately comprehend, develop, and assess this technology efficiently and safely to conquer these challenges.

In this paper, we reviewed the relevant studies and researches done in the context of blockchains' adoption in the healthcare field. Based on these studies, we identified the main gaps and challenges remaining in electronic health records. In our future work, we aim to provide a solution for securing electronic medical records with a blockchain-based system that relies on robust encryption techniques and algorithms to provide a secure and patient-centric ecosystem.

References

1. Almulihi, A., Alassery, F., Khan, A., Shukla, S., Gupta, B., Kumar, R.: Analyzing the implications of healthcare data breaches through computational technique. In: IASC, vol. 32, pp. 1763–1779 (2021). https://doi.org/10.32604/iasc.2022.023460

2. Gahi, Y., El Alaoui, I.: Machine learning and deep learning models for big data issues. In: Maleh, Y., Shojafar, M., Alazab, M., Baddi, Y. (eds.) Machine Intelligence and Big Data Analytics for Cybersecurity Applications. SCI, vol. 919, pp. 29–49. Springer, Cham (2021). https://doi.org/10.1007/978-3-030-57024-8_2
3. Gahi, Y., Guennoun, M., Guennoun, Z., El-Khatib, K.: Encrypted processes for oblivious data retrieval. In: 2011 International Conference for Internet Technology and Secured Transactions, pp. 514–518 (2011)
4. Nakamoto, S.: Bitcoin: A Peer-to-Peer Electronic Cash System (2009)
5. Tanwar, S.: Blockchain revolution from 1.0 to 5.0: technological perspective. In: Tanwar, S. (ed.) Blockchain Technology: From Theory to Practice, pp. 43–61. Springer, Singapore (2022). https://doi.org/10.1007/978-981-19-1488-1_2
6. Efanov, D., Roschin, P.: The all-pervasiveness of the blockchain technology. Procedia Comput. Sci. **123**, 116–121 (2018). https://doi.org/10.1016/j.procs.2018.01.019
7. Blockchain [Book]. https://www.oreilly.com/library/view/blockchain/9781491920480/. Accessed 25 Dec 2022
8. Bhutta, M.N.M., et al.: A survey on blockchain technology: evolution architecture and security. IEEE Access. **9**, 61048–61073 (2021). https://doi.org/10.1109/ACCESS.2021.3072849
9. Venkatesan, S., Sahai, S., Shukla, S.K., Singh, J.: Secure and decentralized management of health records. In: Namasudra, S., Deka, G.C. (eds.) Applications of Blockchain in Healthcare. SBD, vol. 83, pp. 115–139. Springer, Singapore (2021). https://doi.org/10.1007/978-981-15-9547-9_5
10. Kumar Sharma, D., Sreenivasa Chakravarthi, D., Ara Shaikh, A., Al Ayub Ahmed, A., Jaiswal, S., Naved, M.: The aspect of vast data management problem in healthcare sector and implementation of cloud computing technique. Mater. Today Proc. (2021). https://doi.org/10.1016/j.matpr.2021.07.388
11. Haleem, A., Javaid, M., Singh, R.P., Suman, R., Rab, S.: Blockchain technology applications in healthcare: an overview. Int. J. Intell. Netw. **2**, 130–139 (2021). https://doi.org/10.1016/j.ijin.2021.09.005
12. Qiao, R., Luo, X.-Y., Zhu, S.-F., Liu, A.-D., Yan, X.-Q., Wang, Q.-X.: Dynamic autonomous cross consortium chain mechanism in e-healthcare. IEEE J. Biomed. Health Inform. **24**, 2157–2168 (2020). https://doi.org/10.1109/JBHI.2019.2963437
13. Gahi, Y., Alaoui, I.E.: A secure multi-user database-as-a-service approach for cloud computing privacy. Procedia Comput. Sci. **160**, 811–818 (2019). https://doi.org/10.1016/j.procs.2019.11.006
14. Faroukhi, A.Z., El Alaoui, I., Gahi, Y., Amine, A.: A multi-layer big data value chain approach for security issues. Procedia Comput. Sci. **175**, 737–744 (2020). https://doi.org/10.1016/j.procs.2020.07.109
15. Tranfield, D., Denyer, D., Smart, P.: Towards a methodology for developing evidence-informed management knowledge by means of systematic review. Br. J. Manag. **14**, 207–222 (2003). https://doi.org/10.1111/1467-8551.00375
16. El Alaoui, I., Gahi, Y., Messoussi, R., Todoskoff, A., Kobi, A.: Big data analytics: a comparison of tools and applications. In: Ben Ahmed, M., Boudhir, A.A. (eds.) SCAMS 2017. LNNS, vol. 37, pp. 587–601. Springer, Cham (2018). https://doi.org/10.1007/978-3-319-74500-8_54
17. Madine, M.M., et al.: Blockchain for giving patients control over their medical records. IEEE Access. **8**, 193102–193115 (2020). https://doi.org/10.1109/ACCESS.2020.3032553
18. Abdellatif, A.A., Al-Marridi, A.Z., Mohamed, A., Erbad, A., Chiasserini, C.F., Refaey, A.: ssHealth: toward secure, blockchain-enabled healthcare systems. IEEE Network **34**, 312–319 (2020). https://doi.org/10.1109/MNET.011.1900553

19. Mohey Eldin, A., Hossny, E., Wassif, K., Omara, F.A.: Federated blockchain system (FBS) for the healthcare industry. Sci. Rep. **13**, 2569 (2023). https://doi.org/10.1038/s41598-023-29813-4
20. Patientory | Your Health At Your Fingertips. https://patientory.com/. Accessed 25 Dec 2022
21. Hathaliya, J., Sharma, P., Tanwar, S., Gupta, R.: Blockchain-based remote patient monitoring in healthcare 4.0. In: 2019 IEEE 9th International Conference on Advanced Computing (IACC), pp. 87–91 (2019). https://doi.org/10.1109/IACC48062.2019.8971593
22. Tanwar, S., Parekh, K., Evans, R.: Blockchain-based electronic healthcare record system for healthcare 4.0 applications. J. Inf. Secur. Appl. **50**, 102407 (2020). https://doi.org/10.1016/j.jisa.2019.102407
23. Zoughalian, K., Marchang, J., Ghita, B.: A blockchain secured pharmaceutical distribution system to fight counterfeiting. Int. J. Environ. Res. Public Health **19**, 4091 (2022). https://doi.org/10.3390/ijerph19074091
24. Bryatov, S., Borodinov, A.: Blockchain technology in the pharmaceutical supply chain: researching a business model based on Hyperledger Fabric. In: Proceedings of the V International conference Information Technology and Nanotechnology 2019, pp. 134–140. IP Zaitsev V.D. (2019). https://doi.org/10.18287/1613-0073-2019-2416-134-140
25. Chelladurai, M.U., Pandian, S., Ramasamy, Dr.K.: A blockchain based patient centric electronic health record storage and integrity management for e-Health systems. Health Policy Technol. **10**, 100513 (2021). https://doi.org/10.1016/j.hlpt.2021.100513
26. Xiao, Y., Xu, B., Jiang, W., Wu, Y.: The HealthChain blockchain for electronic health records: development study. J. Med. Internet Res. **23**, e13556 (2021). https://doi.org/10.2196/13556
27. Thenmozhi, M., Ranganayakulu, D., Geetha, S., Valli, R.: Implementing blockchain technologies for health insurance claim processing in hospitals. Mater. Today Proc. (2021). https://doi.org/10.1016/j.matpr.2021.02.776
28. Healthcare Data Breach Statistics. https://www.hipaajournal.com/healthcare-data-breach-statistics/. Accessed 13 June 2023
29. Healthcare blockchain adoption rate worldwide 2017. https://www.statista.com/statistics/759208/healthcare-blockchain-adoption-rate-in-health-apps-worldwide/. Accessed 13 June 2023
30. Medtronic: Meaningful Innovation - The Science Behind Healthcare, https://www.medtronic.com/za-en/transforming-healthcare/meaningful-innovation/science-behind-healthcare.html, last accessed 2023/06/13
31. Mazlan, A., Mohd Daud, S., Mohd Sam, S., Abas, H., Rasid, S., Yusof, F.: Scalability challenges in healthcare blockchain system - a systematic review. IEEE Access **8**, 1 (2020). https://doi.org/10.1109/ACCESS.2020.2969230

Design of a Tokenized Blockchain Architecture for Tracking Trade in the Global Defense Market

Mustafa Sanli(✉)

Aselsan, Ankara, Turkey
msanli@aselsan.com.tr

Abstract. The trades in global defense market usually involve sensitive systems, firearms and ammunition which need special regulations and control. Controlling the flow of defense products requires a comprehensive and coordinated effort involving cooperation between manufacturers, local dealers and national law enforcement. However, ensuring transparency in the trade of goods through the defense market is difficult. Traditional approaches can be replaced with a blockchain-based ledger to solve their transparency issues. Blockchain technology and tokenization can be used for making defense sales more transparent and traceable. In this study, NFTs are utilized to build a tokenized architecture that use the Ethereum blockchain, IPFS and smart contracts to trace, perform transactions, and create digital twin certificates for the products of the defense market. Our solution offers security, high integrity, reliability, traceability, and transparency while eliminating any requirements for trusted intermediaries.

Keywords: Blockchain · Tokenization · NFT · Defense Market

1 Introduction

The products traded in the defense market include sensitive systems, firearms and ammunition whose origin and path of delivery are required to be tracked closely. Government authorizations and end user certifications (EUC), which attest that the buyer is the materials' intended final recipient and has no intention of transferring the materials to another, are typically required for the export of sensitive defense systems. Despite the efforts of many governments to restrict the flow of these systems to undesirable locations, there are a number of issues with EUCs as a method of preventing unwanted defense exports. The recipient may not truly keep its word to not transfer the systems received, or EUCs may be forged. EUCs that lack adequate end-use monitoring are seen as ineffective barriers to unauthorized acquisition of defense systems.

Another significant threat to safety and security is the illegal trade in firearms. Conflicts intensify as a result of its expansion, which also makes other criminal acts easier. Finding the connections between firearm smuggling and other forms of organized crime, examining how firearms get into black markets, and figuring out intra- and cross-border flows are all important in the fight against firearm smuggling. As a result, the fight against illicit trade in firearms is a very difficult task. Regulators and governments must make

M. Younas et al. (Eds.): Deep-BDB 2023, LNNS 768, pp. 30–40, 2023.
https://doi.org/10.1007/978-3-031-42317-8_3

monitoring and preventing the illegal firearms trade their top priority in order to combat this.

Ammunition is another factor that needs to be carefully addressed when regulating the defense market. A criminal network needs constant access to ammunition. The kind of firearm used is also determined by the brand and model of ammunition available. Both government arsenals and private sources of ammunition are available. Due to the fact that the majority of lost ammunition is not discovered and recorded, it is challenging to determine how much was lost. In Guanajuato, Mexico, in June 2021, 7 million rounds headed for Texas were stolen [1]. More than 9.5 million rounds of ammunition were lost in South Africa over the course of the preceding six fiscal years, it was revealed in August 2019 [2]. One solution to this issue would be to mandate that ammunition be stamped with a head stamp by manufacturers, just like firearms are. Brazil's practice of this serves as an illustration. Although ammunition is acknowledged as one of the materials that governments must control under the UNTOC Firearms Protocol, no particular steps are needed to register or restrict its sale [3].

The proliferation of additional criminal markets and the escalation of violence are both fueled by the illegal trade in firearms and ammunition. Controlling the flow of firearms requires a comprehensive and coordinated effort involving cooperation between manufacturers, local dealers and national law enforcement. All legally made products must be registered in a global registry that keeps track of licenses and sales. All manufacturers are required to produce in compliance with this record, and unlicensed weapons, including those of craft production, may be seized and disabled.

However, ensuring transparency in the trade of goods for the defense market is difficult. Traditional approaches can be replaced with a blockchain-based ledger to solve their transparency issues. Blockchain works by keeping the list of transactions in blocks. Each block contains a header and a body. The header contains the hash of the previous block, thus linking the blocks together and forming a chain. This method provides an immutability feature that makes it difficult to tamper with data stored in the blockchain [4]. The blockchain is also decentralized because there is no trusted central authority and no need for a reliable third party to verify transactions. One way of performing transactions on a blockchain is implementing smart contracts [5]. Smart contracts are self-executing programs that track and manage transactions. Blockchain technology has a wide range of uses as a secure data storage solution. Examples of these application fields include autonomous vehicle management, healthcare applications, supply chain management, and waste treatment management [6–9]. The immutable nature of the blockchain ledger helps keep an unchanging record of all process flow in the defense market transactions and limits fraudulent activity. When purchasing defense equipment, customers and organizations can use blockchain to track every step and process, generating a more traceable record.

Tokenization, utilizing Non-Fungable Tokens (NFTs) can be used to create digital assets and certificates that identify physical units when building the supply chain for the defense market on the blockchain. NFTs are non-fungible, meaning each coin is unique and cannot be exchanged for another [10]. Each NFT has a single owner and is backed by the blockchain, which adds value and traceability to the ownership of the digital asset. Many applications make use of NFTs by offering digital assets of their physical counterparts that can be used in the virtual world [11]. Similar to this, NFTs can be helpful in codifying the properties of weapons and their subsystems, offering identification and quality assurance, and ensuring a legitimate part purchase, aiding customers in avoiding purchasing weapons from illegal organizations.

In Ethereum, there are three widely accepted standards used in tokenization: ERC-20, ERC-721 and ERC-1155. ERC-721 offers the option of generating unique token IDs, a feature not found in ERC-20. As a result, tokens produced using the ERC-721 standard are commonly referred to as NFTs, whereas tokens produced using the ERC-20 standard are referred to as fungible tokens. Another token standard on the Ethereum blockchain that enables the production of both fungible and non-fungible tokens is ERC-1155, an upgraded standard over ERC-721. The objective is to develop a smart contract interface that can represent both types. Despite the fact that NFTs offer a wide range of applications, from digital art to decentralized gaming, in our design we employ NFTs to produce digital assets for the defense market's products. Each NFT token contains metadata that may include details on the related physical asset, such as the object's production date, the manufacturer, and a variety of other attributes. Images or videos of the products could be included in other metadata. Large files of the metadata require expensive storage on the ledger. Therefore, the IPFS decentralized storage system is used. The IPFS storage is accessed to obtain a URL link that is then used to refer to the NFT metadata.

Decentralized applications (dApp) are utilized to control NFT ownership and transfer. A dApp is a computer program running on the blockchain. dApps basically consist of two components: A front-end user interface, which is usually a web page, and a back-end code that runs on the blockchain in the form of a smart contract. Smart contracts contain the business logic of dApps. They consist of a set of data and functions located at a specific address on the blockchain. They enable writing applications where users can define rules to be executed when pre-defined conditions are met [5].

Integrating blockchain, NFTs, and smart contracts can establish a trustworthy network for managing sensitive defense market assets. This paper proposes a decentralized solution to issue digital certificates for sensitive products in the defense market, and guarantee their delivery throughout the supply chain. Our proposed approach utilizes Ethereum blockchain network, in which our developed smart contracts operate to manage proof of ownership and proof of delivery for sensitive assets. We implemented NFT ownership and transfer management using standard ERC-721 interfaces in the smart contracts. The primary contributions of our work can be summarized as follows:

- We propose an architecture that makes use of NFTs and blockchain to provide proof of ownership and proof of delivery for sensitive products in the defense market.
- We use an ERC-721 based tokenization approach to build a decentralized solution to track and record the flow of products.
- We offer an overview of the system architecture, identify the system stakeholders, and use sequence diagrams to demonstrate the efficient operation of our proposed approach.
- We implement our proposed architecture into smart contracts based on the Ethereum platform and deploy them on various blockchains for testing.

The remainder of our paper is organized as follows. Section 2 provides a brief overview of the related work in this field. Section 3 illustrates how blockchain technology can provide value for trades in the defense market. Section 4 presents our proposed architecture. Our conclusions are given in Sect. 5.

2 Related Work

Although the defense market might be viewed as a niche for blockchain-based solutions, supply chains in other sectors have a wide range of applications. This section discusses some research on blockchain-based frameworks or techniques for smart supply chain management.

[12] highlights the promise of blockchain technology in resolving significant supply chain issues, such as speed, cost, risk, and product quality. Additionally, use case examples from international supply chains, like Maersk and Alibaba, are provided. This work does not, however, include a comprehensive framework for a global supply chain.

A traceability monitoring architecture called BRUCHETTA is presented in [13]. IoT devices are used in the proposed architecture to give customers access to product information at each stage of the oil supply chain. With the aid of six procedures, the Hyperledger Fabric blockchain is used to manage the entire supply chain.

[14] presents a use case for managing the tracking system in the agricultural sector using the Ethereum blockchain. The suggested solution tracks and stores soybean data by means of a decentralized file system and smart contracts.

In [15], a new three-step methodology is presented for integrating blockchain into the food supply chain. Following these steps, nine critical characteristics are discovered that can offer a better scenario while improving the effectiveness of policy-making for a food supply chain that is integrated with blockchain technology.

[16] investigated how blockchain technology may enhance supply chain sustainability. The authors looked at the potential benefits of blockchain technology for supply chain sustainability in detail and concluded that doing so might greatly increase the sustainability of supply chain management systems.

The adoption of blockchain in the pharmaceutical supply chain was examined in [17]. The authors of this study looked into two private blockchains, Hyperledger Fabric and Hyperledger BESU, which can be utilized to create a system for enhancing supply chain traceability and security. They analyzed the two platforms and came to the conclusion that traceability issues can be solved using blockchain technology. They also stressed how crucial it is to consider potential challenges when building this technology.

[18] investigated and suggested the administration of a medical supply chain using blockchain technology. The authors created a solution based on Hyperledger Fabric. They arrived at the conclusion that applying blockchain technology to the supply chain improves medicine transparency, tracing, and tracking.

[19] looked into the automobile industry's need for supply chain integration using blockchain technology. They defined the theory of dynamic capabilities as the capacity of an organization to consciously build, enlarge, and alter its resource base. They followed by describing blockchain as a technology with dynamic capabilities. They performed a survey and received replies from 138 Indian automotive companies, demonstrating that

Blockchain has a beneficial impact on organizations and their resources in supply chain management.

[20] investigated the use of blockchain technology to track pharmaceutical drugs in a supply chain network. The authors of this study used interviews and a questionnaire to figure out the needs and requirements for the new system, and the results were used to ascertain both functional and non-functional requirements. The blockchain-based system, according to the researchers, enhances security, transparency, and traceability.

For improved drug traceability, [21] developed the pharmaceutical supply chain utilizing the Ethereum blockchain. The solidity programming language is employed in the creation of smart contracts. In addition to security and gas cost analyses, this paper provided the system design, algorithms, testing, and validation.

The effectiveness of blockchain technology for traceability in the agricultural food supply chain is evaluated in [22]. They referred to the food supply chain as being the most extensive and complicated, with a lack of transparency. They looked into more traditional approaches to chain traceability, including the usage of RFID tags and a few Internet of Things devices that are connected to a centralized system. Following that, they provided a summary of some advantages of blockchain technology. Upon analysis of blockchain's potential application in the agricultural supply chain, it was found that blockchain can help with data traceability and immutability. Additionally, it promotes decreased risk, cost savings, and time efficiency.

3 Value of Blockchain for Trades in Defense Market

For tracking the defense market products, systems, firearms and ammunition, blockchain technology provides a secure platform to keep immutable records of sensitive information such as production records, delivery logs, usage statistics, certificates and authorizations. As a result of the encryption inherent in the blockchain, the stored data is immune to tampering. Additionally, there is no central authority that controls and regulates the process.

Transparency and monitoring: Data from all parties in the defense supply chain is integrated into a single source of truth through the use of blockchain technology. Every party involved can keep an eye on the product's production and distribution processes as a result of this transparency. Additionally, digital signatures can be utilized for verifying the authenticity of certificates and authorizations. Blockchain technology's transparency makes sure that important credentials like authorizations, certifications, and authenticity are not compromised.

Security and settlement: Blockchain offers a secure and encrypted environment for exchanging data and documents due to its immutable properties achieved through cryptography. Every transaction is recorded in the ledger permanently. Any modifications are also noted and recorded. As a result, the likelihood of deception is decreased.

Decentralization: Decentralized platforms don't have a single point of failure. They can therefore withstand security threats.

Decreased process complexity: Blockchain technology eliminates the need for middlemen by fostering ecosystem trust and enabling peer-to-peer business models. Through the use of smart contracts, blockchain technology automates tasks. The benefits of this automation include the ability to control document flow and organize the order of actions with a visible ledger. Every phase of the process gets recorded on the blockchain when the contract conditions are fulfilled, and the terms and conditions of the smart contracts are completely integrated and visible to all participants. Blockchain further decreases the possibility of human mistake and smart contracts help automate activities.

Data integrity: Small sized data is stored directly on blockchain. The Ethereum network maintains the integrity of all data recorded on the blockchain. Only hash values are kept on the blockchain for larger data. The integrity of data saved outside of the blockchain can be checked using these hash values.

Data confidentiality: Despite the fact that blockchain does not by default provide confidentiality for data kept on the ledger, confidentiality may still be offered given the following two reasons: First, since Ethereum entities are pseudonyms, even if the data is made public, it is difficult to connect them to specific individuals. Second, it is possible to encrypt data before storing on blockchain.

Availability: Blockchain is especially resistant to attacks that would stop actual users from accessing resources because of its decentralized structure. All operations, including minting NFTs and transferring ownership, continue to operate even during threats like denial of service.

Accountability: In our suggested approach, participant tracking is essential to preserving a trustworthy environment. As part of the tamper-proof ledger, the creator and miner of transactions are logged with timestamps. As a result, our approach enables us to monitor and trace all activity. Additionally, by adopting the ERC-721 standard, we make sure that our system has a strong mechanism in place to prevent unauthorized parties from controlling and nefariously moving NFTs.

4 Design and Implementation of a NFT Based Decentralized Architecture

In this section, we present a tokenized architecture for making sales in the defense market more transparent and traceable. Our proposed architecture uses the Ethereum blockchain, NFTs, IPFS and smart contracts to trace, perform transactions, and create digital certificates for the products of the defense market.

We use the Ethereum blockchain to take advantage of its smart contract support and programmability. Additionally, our solution offers security, high integrity, reliability, traceability, and transparency while eliminating any requirements for trusted centralized authority. Contrary to conventional methods, the transfer of ownership for defense market products is completed automatically and securely without the necessity of visiting any regulatory bodies.

4.1 System Design

The conceptual design of our proposed architecture is shown in Fig. 1. Each user of the system has a combination of public and private keys that are used to digitally sign and validate transactions. Additionally, they have a unique address that is created by hashing the user's public key. The system primarily has 5 actors involved.

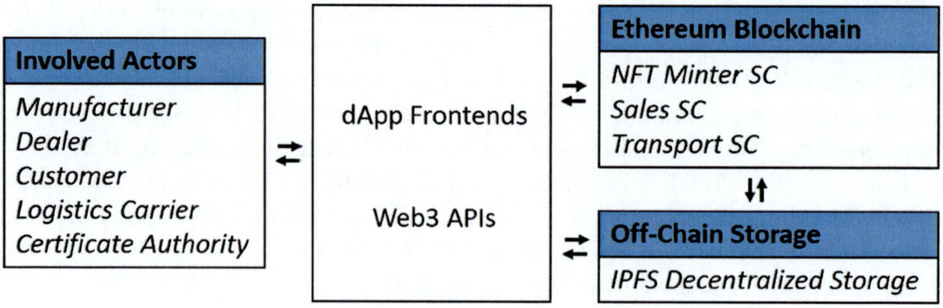

Fig. 1. Conceptual design of the proposed architecture

- Manufacturer: Produces defense products. The products' characteristics and attributes are used to create metadata after production. Additionally, the metadata can include videos or pictures of the product or its component parts. The metadata is uploaded to IPFS storage. A URI (Uniform Resource Identifier) link is received from the IPFS storage. The manufacturer sends the URI to the certificate authority to register it and generate a digital twin certificate using NFTs.
- Dealer: Purchases the products from the manufacturer. Following the purchase, the ownership of the digital twin NFT is transferred to the dealer.
- Certificate Authority: Generates digital certificates by building a digital twin of the product using NFTs. The NFTs include manufacturer address and metadata URI from IPFS storage. After the NFT is minted, the physical item can be transferred to a dealer together with its digital twin NFT.
- Customer: Purchases the products from the dealer. After the purchase, NFT ownership is transferred to the customer.
- Logistics Carrier: Carries the product from dealer to customer. During the transportation period, ownership of the NFT remains with the dealer.

Each actor interacts with the blockchain and the smart contracts through dApp frontends and Web3 APIs. There are three smart contracts in the proposed architecture.

- NFT Minter Smart Contract: Utilizes the ERC-721 library's standard interfaces for managing and minting NFTs. Only the certificate authority is allowed to call the minting function. The certificate authority provides a metadata URI with product attributes. Following minting, an event is set off to inform all network actors that a new digital twin has been certified.
- Sales Smart Contract: Products may be listed for sale by the dealer. An event called a transport request is created after the selling party approves the purchase.
- Transport Smart Contract: Supervises the coordination of product delivery between the dealer, and the customer. Following the request event that the Purchase Smart Contract generated, the delivery is started. Both transmitting and receiving parties make security deposits to guarantee policy compliance. Deposits are reimbursed to the parties following the transfer.

The sequence diagrams illustrating the interactions of the actors in the system are given in Figs. 1, 2 and 3. Our smart contract code is also provided for reference in [37] (Fig. 4).

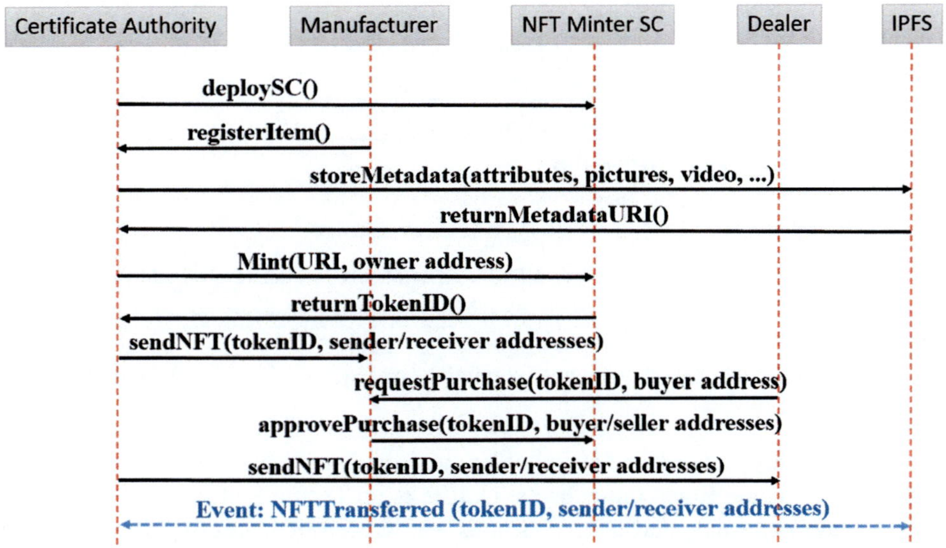

Fig. 2. Sequence diagram showing the interactions of actors following the manufacturing of a product

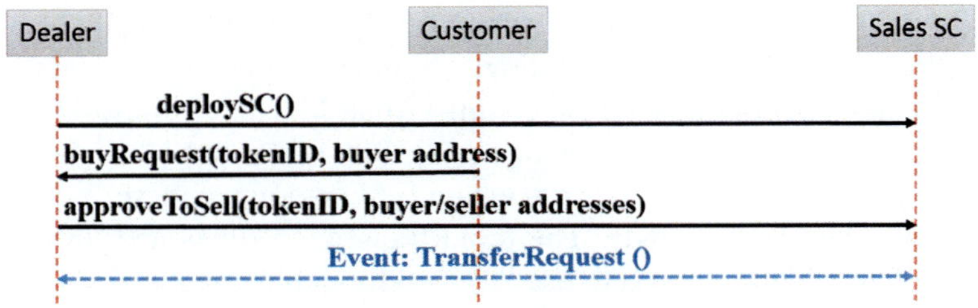

Fig. 3. Sequence diagram showing the interactions of actors during the sale of a product

4.2 Implementation and Testing

Our proposed architecture has been designed and implemented on Ethereum Ropsten [23] and Polygon Blockchains [24]. The Solidity Programming Language [25] and Remix Integrated Development Environment [26] are used to create the smart contracts of the project. All the functions contained in the smart contracts were extensively tested to verify that the defined rules and functionalities of the smart contracts were not violated in a practical process execution. Modifiers are used in both contracts to make sure that only authorized parties or system users are permitted to carry out a function. A number of events are also used to track and validate the logs of transactions that

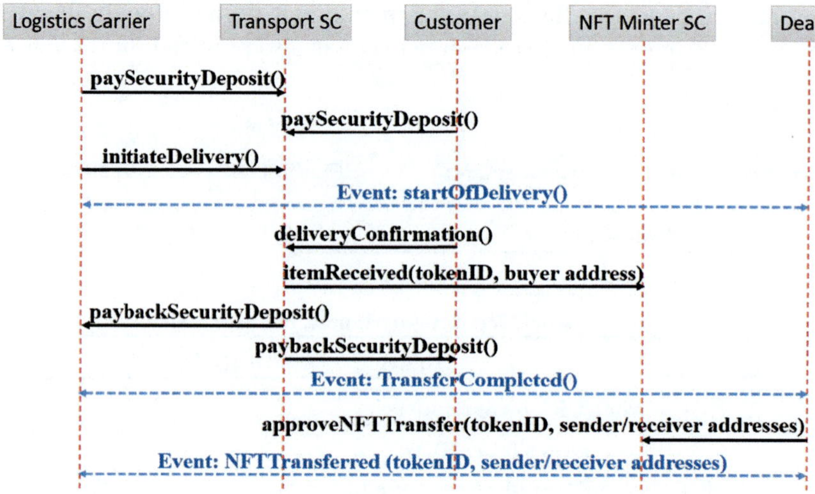

Fig. 4. Sequence diagram showing the interactions of actors during the transport of a product

have been carried out. All the stakeholders and the smart contracts are identified with a uniquely assigned Ethereum address. Unit tests on Truffle Suite [27] are run in order to test smart contracts, and test conditions are written in Javascript. Behavioral tests are carried out on the local Ganache Blockchain before contracts are deployed to the Ethereum Ropsten Blockchain [28]. In the user interface, Metamask Browser Extension [29] is used to access the blockchain via Ethereum-compatible JSON RPC Application Programming Interface (API) [30]. Web3 Javascript Library [31] enables users to communicate with the blockchain using a web page. Contracts have finally been put on the Polygon Blockchain after complete testing. Although Polygon Blockchain was selected as the production environment, all Ethereum Virtual Machine compatible blockchains, including Ethereum Mainnet and Testnets [32], Binance Smart Chain [33], Avalanche C-Chain [34], TOMO Chain [35], Tron Chain [36], and many more, can run the smart contracts developed in this study. The rapid transaction times and low fees that this blockchain offers are the reasons to use Polygon for deployment.

5 Conclusion

Trades in the defense market entail goods whose origin and route of delivery must be strictly monitored. Tokenization and blockchain technology can be used to increase the transparency and traceability of defense sales. In this study, NFTs are used to establish a tokenized architecture that makes use of the Ethereum blockchain, IPFS, and smart contracts to track, carry out transactions, and provide digital twin certifications for defense industry items. The Polygon Blockchain is used to implement the suggested architecture. Our solution offers security, high integrity, reliability, traceability, and transparency while eliminating any requirements for trusted centralized authority. Contrary to conventional methods, the transfer of ownership for products used in the defense market is accomplished automatically and securely without the need of completing challenging regulatory paperwork.

References

1. Arms Trade Treaty. https://thearmstradetreaty.org/treaty-text.html. Accessed 01 May 2023
2. del Bajío, E.S.: https://www.elsoldelbajio.com.mx/policiaca/investigan-a-cusaem-por-robo-de-7-millones-de-municiones-asalto-a-cargamento-de-municiones-fue-realizado-por-cartel-6851532.html. Accessed 01 May 2023
3. The firearms protocol. https://www.unodc.org/unodc/en/firearms-protocol/the-firearms-protocol.html. Accessed 01 May 2023
4. Zheng, Z., Xie, S., Dai, H., Chen, X., Wang, H.: An overview of blockchain technology: architecture, consensus, and future trends. In: International Congress on Big Data (BigData Congress), IEEE (2017)
5. Szabo, N.: The idea of smart contracts. Nick Szabo's Papers and Concise Tutorials, vol. 6 (1997)
6. Gao, W., Hatcher, W.G., Yu, W.: A survey of blockchain: techniques applications and challenges. In: International Conference on Computer Communication and Networks (ICCCN), IEEE (2018)
7. Kaushik, A., Choudhary, A., Ektare, C., Thomas, D., Akram, S.: Blockchain - literature survey. In: International Conference on Recent Trends in Electronics Information & Communication Technology (RTEICT), IEEE (2017)
8. Tasatanattakool P., Techapanupreeda C.: Blockchain: challenges and applications. In: International Conference on Information Networking (ICOIN), IEEE (2018)
9. Abbas, E., Sung-Bong, J.: A survey of blockchain and its applications. In: International Conference on Artificial Intelligence in Information and Communication (ICAIIC), IEEE (2019)
10. Antonopoulos, A.M., Wood, G.: Mastering Ethereum. O'Reilly Media, Sebastopol, CA (2019)
11. Rehman, W., Zainab, H.E., Imran, J., Bawany, N.Z.: NFTs: applications and challenges. In: International Arab Conference on Information Technology (ACIT), IEEE (2021)
12. Kshetri, N.: Blockchain's roles in meeting key supply chain management objectives. In: International Journal of Information Management, IEEE (2018)
13. Arena, A., Bianchini, A., Perazzo, P., Vallati, C., Dini, G.: Bruschetta: an IoT Blockchain-based framework for certifying extra virgin olive oil supply chain. In: International Conference on Smart Computing (SMARTCOMP), IEEE (2019)
14. Salah, K., Nizamuddin, N., Jayaraman, R., Omar, M.: Blockchain-based soybean traceability in agricultural supply chain. IEEE Access, vol. 7 (2019)
15. Khan, U., An, Z.Y., Imran, A.: A blockchain ethereum technology-enabled digital content: development of trading and sharing economy data. IEEE Access (2020)
16. Park, A, Li, H.: The effect of blockchain technology on supply chain sustainability performances. Sustainability (2021)
17. Uddin, M., Salah, K., Jayaraman, R., Pesic, S., Ellahham, S: Blockchain for drug traceability: architectures and open challenges. Health Inf. J. **27**, 146045822110112 (2021)
18. Lingayat, V., Pardikar, I., Yewalekar, S., Khachane, S., Pande, S.: Securing pharmaceutical supply chain using Blockchain technology. In: ITM Web of Conferences (2021)
19. Kamble, S.S., Gunasekaran, A., Subramanian, N., Ghadge, A., Belhadi, A., Venkatesh, M.: Blockchain technology's impact on supply chain integration and sustainable supply chain performance: evidence from the automotive industry. Ann. Oper. Res. **327**, 575–600 (2021)
20. Mohana, M., Ong, G., Ern, T.: Implementation of pharmaceutical drug traceability using blockchain technology. INTI J. **35** (2019)
21. Musamih, A., Salah, K., Jayaraman, R., Arshad, J., Debe, M., al-Hammadi, Y., Ellahham, S.: A Blockchain based approach for drug traceability in healthcare supply chain. IEEE Access, **9**, 9728–9743 (2021)

22. Demestichas, K., Peppes, N., Alexakis, T., Adamopoulou, E.: Blockchain in agriculture traceability systems: a review. Appl. Sci. **10**, 4113 (2020)
23. Ropsten Blockchain. http://ropsten.etherscan.io. Accessed 01 May 2023
24. Polygon Blockchain. http://polygon.technology. Accessed 01 May 2023
25. Solidity. http://soliditylang.org. Accessed 01 May 2023
26. Remix IDE. http://remix.ethereum.org. Accessed 01 May 2023
27. Truffle Suite. http://trufflesuite.com. Accessed 01 May 2023
28. Ganache Blockchain. http://trufflesuite.com/ganache. Accessed 01 May 2023
29. Metamask. http://metamask.io. Accessed 01 May 2023
30. JSON RPC API. http://ethereum.org/en/developers/docs/apis/json-rpc. Accessed 01 May 2023
31. Web3 Javascript Library. http://ethereum.org/en/developers/docs/apis/javascript. Accessed 01 May 2023
32. Ethereum Blockchain. http://ethereum.org. Accessed 01 May 2023
33. Binance Smart Chain. http://www.binance.org/smartChain. Accessed 01 May 2023
34. Avalanche Blockchain. http://www.avalanche.network. Accessed 01 May 2023
35. Tomo Blockchain. http://tomochain.com. Accessed 01 May 2023
36. Tron Blockchain. http://tron.network. Accessed 01 May 2023
37. Smart Contract Code. http://www.mustafasanli.com/sc. Accessed 01 May 2023

Requirements for Interoperable Blockchain Systems: A Systematic Literature Review

Senate Sylvia Mafike and Tendani Mawela(✉)

University of Pretoria, Hatfield 0002, Pretoria, South Africa
senatemafike@gmail.com, tendani.mawela@up.ac.za

Abstract. The lack of standards or methods guiding the development of blockchain systems has led to the existence of heterogeneous blockchains that are unable to communicate. As a result, organizations using the technology are unable to collaborate and share data with their counterparts. Extant literature discussions focus predominantly on the mechanisms used to achieve interoperability and neglect the requirements needed to achieve blockchain interoperability. This paper aims to address the blockchain interoperability challenge by exploring the fundamental requirements that blockchain systems should offer to achieve technical, semantic, legal and organizational interoperability. The study adopted the Systematic Literature Review approach and included 83 peer-reviewed articles and 5 industry reports on blockchain interoperability. The findings of the study are categorized according to the four levels of interoperability defined by the European Interoperability Framework. The findings indicate security and data confidentiality as some of the technical requirements that must be met. While semantic requirements include establishing standardized data ontologies. Legal interoperability requirements include the use of smart contracts to enforce domestic and international laws, and organizational requirements relate to establishing cooperation agreements and developing collaborative business models for data exchanges between organizations.

Keywords: Blockchain interoperability · Cross-blockchain communication · Cross-blockchain technology · Requirements

1 Introduction

Blockchain is a technology famously known as the foundational technology for cryptocurrencies such as Bitcoin and Ethereum. However, in recent years, its utility has expanded beyond its original use in cryptocurrencies. Now blockchain can be found in enterprises across different industries. Industries such as supply chain [1], health [2], and finance [3], are leveraging the technology to address industry-specific challenges to enforce trust, improve stakeholder communication and collaboration, reduce operational costs, and to generate new business models [4]. The challenge, however, with the use of blockchain in addressing industry-specific challenges is that the resulting industry solutions are not standardized. The lack of standards guiding the development of

M. Younas et al. (Eds.): Deep-BDB 2023, LNNS 768, pp. 41–55, 2023.
https://doi.org/10.1007/978-3-031-42317-8_4

blockchain solutions across industries has led to the emergence of disparate and incompatible blockchain ecosystems which cannot communicate with each other. Differences in the core consensus mechanisms, protocol and smart contract structures [5] inhibit the interoperability of these blockchain ecosystems. This means that blockchain systems amongst and within different industries cannot share data seamlessly even in cases where such collaborative sharing of data is critical.

The concept of interoperability is not new to the Information Systems field. The ever-changing technological landscape means that new technologies are continuously introduced into existing enterprises, sparking the need to integrate these technologies with existing enterprise legacy systems. The interoperability of organizations through their processes and ICT systems is critical in ensuring that there is a seamless and effective collaboration and reduced information asymmetry between and within networked enterprises. Interoperability is defined as "the ability of two or more systems or components to exchange information and to use the information that has been exchanged" [6]. These components can be ICT systems, business processes, organizations and e-Government processes. The European Union defines interoperability as "the ability of organizations to interact towards mutually beneficial goals, involving the sharing of information and knowledge between the organizations, through their business processes and by means of the data exchange between their ICT systems [7].

In the context of blockchain, interoperability represents "the ability to share both digital assets and transaction data across distinct networks, without any reliance or restriction by trusted third-party exchanges" [8]. In essence, blockchain interoperability enables data and value transfer between two separate blockchain networks. The value transfer represents the change of ownership of a digital asset stored on a blockchain network [9]. The data and value transfer can occur on a single blockchain network, across heterogeneous blockchain networks or between a blockchain and non-blockchain system. The transfer of value within the same blockchain follows a simple process in which the transacting users only need to know each other's cryptographic keys (wallet addresses). However, the exchange of value between different networks (cross-blockchain asset transfer) is much more complex. Cross-blockchain transfer involves the moving of digital assets from one blockchain to another, thus allowing the user to hold different denominations of the same asset on different blockchain networks [10]. This process requires that one blockchain makes state changes based on information from another [9]. It requires the communicating blockchain networks to be able to verify each other's state and to provide guarantees of the validity of the transactions on each network [11]. However, validating and verifying the cross-blockchain data and state information is a complicated undertaking. This is because it involves the integration of platforms, consensus mechanisms, protocols, smart contracts, governance structures and information sources that are incompatible by design.

1.1 Research Problem

The complexity associated with integrating and interoperating heterogeneous blockchains means that enterprises cannot fully realize the benefits of utilizing blockchain in their operations. Currently, blockchain platforms do not offer a direct way of transferring data and assets across heterogeneous networks. Users and enterprises have to rely on third-party interoperability solutions to transfer data and assets between blockchain networks. The challenge with this is that these third-party solutions compromise the security and privacy of the information being shared [12]. In addition, recent interoperability solutions are themselves not interoperable [9]. The problem of the lack of interoperability is exacerbated by the absence of well-established standards, models and frameworks that guide the development of blockchain enterprise systems [11]. Existing solutions and studies focus mainly on addressing interoperability from a technical and semantic perspective. However, to achieve 'true interoperability' many other aspects must be considered such as legal agreements, governance structures, data formats, semantic choices, applications, technical infrastructure, privacy and security issues [13]. This implies that fully understanding blockchain interoperability requires extending our focus beyond the technical and semantic interoperability to include other levels of interoperability such as organizational and legal interoperability [7].

To address these gaps, this study performed a systematic review of peer-reviewed and grey literature on blockchain interoperability. The objective was to identify and categorize key requirements for achieving technological, semantic, organizational and legal interoperability for blockchain systems. The following research question was formulated and addressed in this study:

RQ 1: *What are the requirements for technical, semantic, organizational and legal interoperability in blockchain systems?*

1.2 Key Contributions

This study makes the following contributions:

- The study extends the ongoing research on blockchain interoperability by exploring interoperability beyond the technical and semantic focus of existing research to include legal and organizational perspectives on blockchain interoperability.
- We provide a comprehensive review of enterprise blockchain interoperability requirements to consider when designing interoperable blockchain solutions. This has practical implications in that it provides developers and designers with a premise and understanding of what the key requirements are for enabling interoperability capabilities in blockchain systems.

The rest of this paper is organized as follows. Section 2 presents an overview of blockchain interoperability and related studies. Section 3 discusses the methodology and Sect. 4 presents the findings and discussions. The paper is concluded in Sect. 5.

2 Blockchain Interoperability Overview

Blockchain interoperability can present in different forms and serve varying purposes depending on the components that need to be connected and the type of data that needs to be shared. For instance, interoperability capabilities are required to connect homogeneous blockchains with the same block structure and consensus mechanism, in which the target blockchain can understand the transaction on the source blockchain [14]. Alternatively, it can represent interoperability between separate blockchains with different consensus algorithms and governance structures. On the other hand, interoperability can be required between a blockchain and a non-blockchain system [15]. This form is prevalent in enterprises and industries that wish to connect their existing systems to external or internal blockchain systems.

2.1 Related Studies

There has been an increase in the number of studies on blockchain interoperability in the main and several surveys and reviews have been conducted on the topic. A recent survey was presented by [16]. Their work presented a review of current cross-chain interoperability solutions, their benefits and challenges. They highlighted that there is a need for a generic interoperability architecture or model for blockchain interoperability. Similarly, Belchior, Vasconcelos [11] presented a survey of challenges, use cases, technologies and solutions for blockchain interoperability. Similarly, they identified the need for standards, models and frameworks to guide the design and implementation of blockchain systems as an open issue. Others; [15], Qasse, Talib [17, 18] also provide different classifications of existing blockchain interoperability solutions. The above studies provide an understanding of current solutions for achieving blockchain interoperability or cross-chain communication, however, they do not provide any design guidelines to guide organizations in their choice of interoperability solutions. Furthermore, the studies focus on technological and semantic interoperability and do not discuss interoperability requirements and solutions relating to other levels of interoperability such as organizational and legal interoperability.

Only a few studies include interoperability aspects relating to legal and organizational levels. For instance, Pillai, Biswas [9] proposed an integration design decision framework that identifies key security assumptions and characteristics of the cross-blockchain technology. Their framework identifies interoperability requirements and maps these requirements to relevant solution options and security assumptions. Others, Nodehi, Zutshi [5, 19], have proposed frameworks to assist organizations in their choice of blockchain interoperability solutions. Belchior, Riley [19], proposed several frameworks in their work. One of their frameworks is a decision framework to assist organizations in selecting the appropriate integration mechanism to connect their blockchain systems. Unlike the other studies that focused on one type of interoperability, their study also included an interoperability assessment for legal, and organization interoperability. Similarly, Nodehi, Zutshi [5] proposed a generic enterprise blockchain design framework which identifies blockchain use cases and also highlights characteristics and configurations required when designing blockchain systems. The framework also includes an interoperability layer that demonstrates how organizations can connect their blockchain solutions to external non-blockchain systems.

Although the above studies provide invaluable insights regarding the existing blockchain interoperability solutions and the considerations for selecting the appropriate solutions, they are limited in their coverage of other forms of interoperability which need to be considered for enterprise blockchain systems. The existing studies do not define any requirements needed for a blockchain solution to be interoperable. There is therefore a need to explore the fundamental required properties needed to design interoperable blockchain systems.

3 Methodology

To provide a comprehensive, transparent and reproducible study of existing literature on blockchain interoperability requirements, we followed the systematic literature review guidelines suggested by [20] and PRISMA. Both the Kitchenham [20] guidelines and the PRISMA guidelines were adopted to ensure that the process of conducting the SLR is systematic, unbiased and that the results are valid and answer the research question. The PRISMA guidelines also helped to enhance the quality of our reporting. The process includes three steps: planning, conducting and reporting the review. In the planning phase, research questions are formulated and a protocol is developed. Conducting the review step involves identifying key sources for the literature to be reviewed, formulating search strings and searching and screening studies.

Search Strategy: The study included peer-reviewed literature and grey literature in the form of technical and enterprise reports. Grey literature is a valuable data source for research intended for academics and practitioners as it provides trust between academics and practitioners and enhances the applicability of academic research work to industry settings [21]. The peer-reviewed articles were sourced from the following databases: ACM Digital Libraries, IEEEXplore Digital Libraries, Science Direct, and Springer Link whereas the grey literature was sourced through a Google search. Search strings were formulated using the following main keywords from the research questions: blockchain interoperability, requirements, elements, framework, and solution. Synonyms, related terms and abbreviations were also used to compile additional search strings to enhance the search effort. A total of 83 peer-reviewed articles and 5 industry reports were reviewed.

The retrieved literature was subjected to a screening process based on the following inclusion and exclusion criteria. Conference and journal articles as well as industry reports written in English, published from 2009 to 2023 and focusing on blockchain interoperability were included. The screening criteria were used to filter articles during the database search. Studies that met the inclusion criteria during the database search were further screened by filtering out duplicates and reviewing their abstracts. The full-text screening was carried out on the remaining papers. The details of the search process are outlined in the PRISMA chart (Fig. 1).

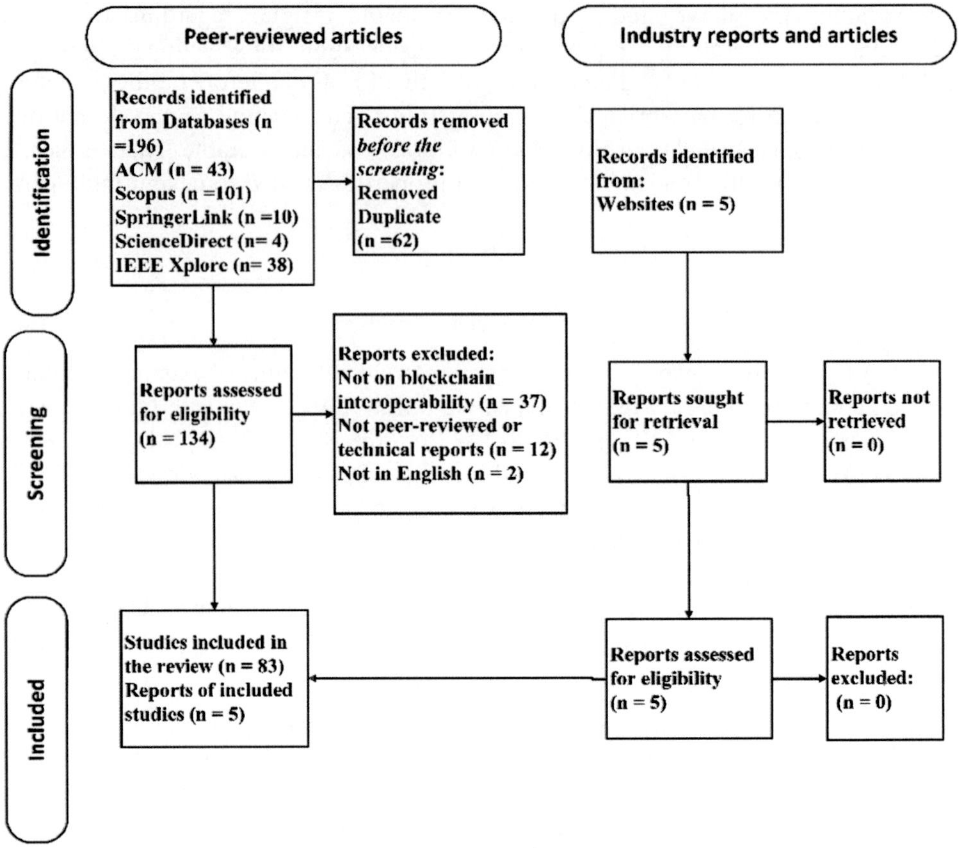

Fig. 1. PRISMA chart representing the search process

4 Results and Discussion

***RQ 1**: What are the requirements for technical, semantic, organizational and legal interoperability in enterprise blockchain systems?*

Achieving interoperability between organizations, their processes, and technology systems, requires meeting specific interoperability requirements which specify what should be in place to enable effective communication and collaboration between entities. To understand the interoperability of blockchain systems we analyzed and reviewed the literature on blockchain interoperability to identify the requirements for achieving interoperability for the four levels of interoperability defined by [7]. Our analysis revealed that the majority of the studies address technical and semantic interoperability as one, as such the requirements specified include both these levels of interoperability. On the other hand, requirements for organizational and legal interoperability were not explicitly stated in most of the reviewed articles. It was also noted that the reviewed literature distinguishes between interoperability requirements for permissioned and permissionless blockchain types. Following this observation, we present the findings on the interoperability requirements and categorize them according to whether they are requirements for permissioned or permissionless blockchain types.

As mentioned in previous sections, blockchains can be either permissioned or permissionless. The different types of blockchains have different interoperability requirements which need to be fulfilled relating to identity, authorization, confidentiality and governance [22].

4.1 Technical and semantic interoperability requirements.

Security, Privacy and Data Confidentiality: Security is a fundamental requirement for interoperability in both permissioned and permissionless blockchains. Both permissioned and permissionless blockchains, need to be able to exchange arbitrary data and digital assets securely. Providing interoperability should not compromise the security of the communicating blockchains. This implies that the mechanism, methods, and operations used to enable interoperability between the blockchains need to be secure. Data exchange between different blockchains should be protected at the source blockchain, when in transit, and at the destination blockchain [23]. This requires both communicating blockchains (permissioned and/or permissionless) to provide measures for ensuring that the data is secure. This means that the source chain should have measures to record and ensure that the data/assets to be transmitted is reliable, data in transit cannot be tampered with, and the destination chain is required to be able to verify and validate the received information [23]. Because interoperating blockchains often requires the use of some integration mechanisms to connect the communicating blockchains, the integration mechanism should also fulfil some security requirements to protect the data in transit.

The integration mechanism needs to be credible and trustworthy. Credibility and trustworthiness are key requirements for integration mechanisms which rely on a central party such as third-party integration modes, single notary schemes and bridges. The integration mechanism should not be less secure than the blockchains it is connecting as this would compromise the security of the connected chains [9]. Furthermore, the integration mechanism needs to be fault tolerant to ensure the continuous availability and accessibility of the data [9].

Concerning data confidentiality and privacy requirements for interoperability in permissioned and permissionless blockchains, permissioned blockchains tend to have stringent requirements for data privacy. Permissioned networks may be used to store confidential information, as a result, connecting permissioned networks should be done in a manner that preserves the confidentiality of the data during cross-chain communication [24]. This is particularly critical for cross-chain communication in which a permissioned blockchain interacts with a permissionless blockchain.

Anonymity Requirements: Permissioned blockchains need to keep the contents of transactions confidential. To fulfil these requirements, the communicating blockchains and the cross-chain technology are required to allow for blockchain identifications and provide authentication mechanisms to enable the digital authentication of the blockchains [25]. On the other hand, in permissionless blockchains, the security and privacy requirements of the data are inherently addressed by decentralization. However, in some cases, different security requirements may be required such as in atomic swaps which may require the cross-chain integration mechanism to fulfil security requirements to ensure the integrity of the information shared [26].

Distinguishability/identifiability: Interoperable blockchains need to be distinguishable or identifiable. Each blockchain needs to have a unique identifier or name for addressing or routing purposes [24]. To achieve this, each interoperable blockchain requires a unique key to identify it during the authentication and routing of exchanged data [27]. For permissioned blockchains specifically, data exchange depends on the ability of the network to authenticate and validate requests and signature proofs accompanying the data [28]. Therefore, in interactions involving permissioned blockchains, the communicating blockchain networks may be required to have and know the identifiers of each other's members [24]. For example, in consortium-type blockchains, all member nodes may be required to be identified and registered [24]. However, there may be differences in how individual networks handle membership identities, thus a cross-network identity management mechanism is required when multiple blockchains interact [28]. The cross-network identity management scheme should adhere to the privacy and security requirements of the network. Furthermore, cross-chain identity management requires: that member identification should be independently verifiable by external entities while still under the owner's control, decentralized identity registers are required to map external identities to network-specific identities, and each network should maintain the integrity of the network's membership and ensure the availability of its membership list to a communication peer during an interoperability session [28]. Furthermore, the digital identity of each blockchain and its members should be verifiable [29]. This means that any blockchain receiving the identity credentials from another blockchain should be able to query the identity register and obtain the identity certificate as proof that the credentials are valid. However, it should be noted that the requirements for identification may vary according to the purpose of the blockchains.

Customization or Standardization of Data: Semantic interoperability requires the standardization of the data formats to enable all participants in the data exchange to verify the reliability of the information [30] and to understand the shared data [31]. For instance, asset exchanges (atomic swaps) require a standardized way of defining asset profiles to ensure that the communicating participants have the same definition of the asset being exchanged [32]. This means that the transacting parties must have an agreement on how the asset value is represented [32].

Standardized Cross-Chain Crotocols: Achieving technical interoperability between disparate blockchains requires that both blockchains should share a standard network communication protocol [33]. Therefore, standard cross-chain protocols should be developed to enable data and asset transfer between different blockchains and to perform value conversions of incoming and outgoing data [32]. In general, cross-chain protocols should fulfil the verification requirement. Meaning that they should enable receiving blockchain to verify the existence and validity of a transaction that occurred on the source blockchain [34]. In addition, cross-chain protocols should fulfil the atomicity and liveliness requirements [35], and double spending prevention and finality [34]. The liveliness requirement ensures that assets are not locked indefinitely and requires the protocol to be synchronous. On the other hand, the atomicity requirement ensures consistency in the states of the blockchains involved in cross-chain communication. Atomicity refers to the notion that all parts of the transfer are committed or the transaction is rolled back. Atomicity ensures that there is no burn without a claim [34]. Furthermore, for permissionless blockchains, cross-chain protocols need an incentive mechanism to encourage good behavior [35] whereas permissioned blockchains may require reputation mechanisms or external enforcement such as legal action. In essence, a reliable cross-chain consensus protocol should support the diversification of blockchains to ensure that minimal modifications to the existing protocol of each system are needed when a new blockchain is introduced to the ecosystem [23].

Platform and Technology Agnostic: Blockchain cross-chain solutions should support communication between heterogeneous blockchains, regardless of the nature of the blockchains (permissioned or permissionless, public or private), the architecture and/or platform [36]. Making the interoperability solution independent of the underlying technology of the blockchains, simplifies the integration process [22].

4.2 Organizational Interoperability Requirements

Interoperability at the organizational level is concerned with the ability of different autonomous organizations and enterprises to cooperate regardless of the differences in practices, culture, legislations and business models [37]. This level is defined by the agreed definitions of technical and semantic interoperability between companies and it includes procedures and rules of how collaborative participation is governed [30]. In terms of blockchain, organizational interoperability is characterized by the use of blockchain technology as a supportive technology to meet the specific operational needs of enterprises in collaborative partnerships. These organizations may deploy different blockchain systems which meet their respective goals, however, interoperability between these systems may be required to facilitate data exchanges between the respective blockchain systems [30]. Organizational interoperability in the context of blockchain systems is the least discussed in the literature as a result we could identify only a few requirements relating to this form of interoperability from the reviewed literature. The identified requirements are discussed as follows:

Business Model Requirements: Blockchain technology can enable organizations to form collaborative relationships that enable them to generate value for themselves and their customers. Through blockchain, organizations can share information across organizational boundaries. To achieve this, organizations are required to have shared collaborative business models for information sharing [38].

Trust Requirements: Establishing inter-organizational collaboration and interoperability requires trust between the parties involved. Trust ensures that business processes are executed correctly. Traditional mechanisms for establishing trust among organizations include mutual agreements and reputation systems. Similarly, for organizations using blockchain technology to collaborate, establishing trust may require cooperation agreements. However, instead of traditional agreements, smart contracts can be used to "enforce the executions of transactions of untrusted stakeholders and ensure the satisfaction of contractual conditions and obligations"[40]referenced in [40].

Governance Requirements: Establishing interoperability at the organizational level means that tasks and processes will be undertaken by multiple parties. Coordinating these tasks requires that the governance models of all stakeholders should be comparable and compatible [40]. However, it is not always possible to have compatible governance models due to the differences in business requirements. Therefore, to achieve effective coordination of tasks and processes, and to enforce trust between the parties requires a composite governance model to be developed to accommodate all the stakeholder [31].

4.3 Legal Interoperability Requirements

Legal interoperability is concerned with "ensuring that organizations that operate under different legal frameworks, policies and strategies are able to share information" [7]. In this study, we consider legal interoperability in the context of blockchain technology. In particular, we consider it to refer to all legal aspects associated with interoperating blockchains, such as local and global legal and regulatory aspects, legalities of smart contracts used to enable cross-blockchain communications and any other legal issues concerning the exchange of data and assets across organizational and jurisdictional boundaries. It should be noted that due to divergences in industry-specific regulations and legal frameworks, the requirements discussed are general legal requirements, and are not specific to any industry or jurisdiction. However, in practice, industry and jurisdiction-specific legal and regulatory requirements would have to be taken into account when developing applications that require interoperability between blockchains.

Transfer of Ownership Legal Requirements: Blockchain interoperability facilitates asset transfers between entities on different blockchains. Asset transfers involve the ownership transfer from one party to another. In permissionless blockchains, the transfer of asset ownership is enforced through underlying features of the technology such as cryptographic keys and consensus mechanisms. However, Ownership transfer, particularly for complex asset tokens like property, has legal implications that may need to be fulfilled during the exchange. Blockchain-based asset transfer mechanisms address double spending but do not take into account that there may be mistakes in some asset transfers and that some asset transfers may be fraudulent [41]. Erroneous and fraudulent asset transfers have legal implications which should be addressed. Therefore protocols or mechanisms should take into consideration erroneous or fraudulent transfers [41]. For instance, in a case where a sender transfers more funds than intended, a traditional blockchain asset transfer mechanism may still validate and effect this transaction, even though legally this transaction is void. Addressing these challenges requires legal controls and agreements to govern blockchain-based asset transfer mechanisms [41]. This could be achieved by embedding a relevant choice of law in the blockchain protocol and through adequate design solutions that allow for a mapping of the smart contract and the paper counterpart [42].

Identification Requirements: Although privacy and anonymity are key features of blockchain technology, these requirements may have to be compromised for legal compliance. In some circumstances, identifying participants in a cross-chain transaction may be required to ensure legal compliance. Some national contract laws may require parties involved in a contractual agreement to be identifiable [42]. For instance, in the case where asset exchange is a payment for services provided, the parties involved in the transaction may need to be identified for tax compliance purposes. This is also the case for permissioned blockchains where the users may be required to authenticate themselves to comply with certain legal regulations such as the Know your customer regulations [8]. Identification may also be required to comply with Anti Money Laundering (AML) and Countering Financial Terrorism (CFT) regulation requirements for asset transfer and exchanges [26].

Jurisdictional Requirements: Blockchain interoperability is not limited to connecting blockchains in the same industry or locality, but can also enable blockchains that span different geographical locations to connect. Different locations may have varying laws. In circumstances where collaboration is required between blockchain systems in different jurisdictions, the systems and parties involved in the data and asset exchanges may have to comply with different laws. In this case, the interoperability mechanisms and smart contracts used are required to anticipate the jurisdictional variations and to include policies and legal controls to address the jurisdictional uncertainties [26, 42].

Smart Contract Requirements: As mentioned above, smart contracts can be used to implement some of the legal requirements needed to achieve legal interoperability. However, for smart contracts to be legally enforceable, they "must satisfy the relevant validity requirements in domestic contract law" [42].

Collaboration Agreements Requirements: Achieving interoperability between multiple parties requires a collaborative effort from all parties involved. Establishing these collaborations has legal implications such as the need to develop cooperation agreements [5]. The agreements serve to identify the role and level of contribution of each party to avoid identifying the level of contributions and cooperation of each stakeholder. A summary of the identified requirements is shown in (Table 1).

Table 1. Summary of the requirements

Interoperability level	Requirements	Supporting Reference
Technical & Semantic	TS-Req1 – security, privacy, confidentiality, credibility and trustworthiness TS-Req 2 - Identifiability of participating blockchains TS-Req 3 - authentication mechanism for participants (e.g. access control) TS-Req4 – Standardization of data models and ontologies TS-Req 5 – verifiability TS-Req6 - liveliness TS-Req 7 – atomicity of transactions TS-Req 8 – Reliability TS-Req 9- decentralization TS-Req 10 – technology and platform agnostic	[9, 23] [24, 27–29] [24, 35] [30–32] [34] [35] [23] [22, 36]
Organizational	O-Req 1 – blockchain-driven business models for data exchange O-Req 2 – Trust requirements O-Req 3 – Governance	[38] [40] [31, 40]
Legal	L-Req 1- Jurisdictional law requirements L-Req2-Identification/privacy requirement L-Req 3 – transfer of ownership L-Req 4 – Smart contracts L-Req 5 - Collaboration agreements	[26, 42] [8, 26, 42] [41, 42] [42] [5]

5 Conclusion

This study aimed to identify interoperability requirements of blockchain systems. To achieve this, a systematic review of 83 peer-reviewed and 5 grey literature papers was carried out. The requirements were categorized according to the four levels of interoperability defined in the European Interoperability framework mainly technical, semantic, legal, and organizational levels. Our analysis found that requirements vary depending on the context and business application of the technology. However, there are fundamental requirements that must be met for blockchain systems to be interoperable. We identified that security and reliability are some of the basic technical requirements, while standardized data formats and ontologies and the requirements for semantic interoperability. It was observed that the majority of the literature does not distinguish between technical and semantic interoperability. In addition, legal and organizational requirements were the least discussed in the literature.

This paper contributes to the existing work on blockchain interoperability by exploring the underexplored area concerning interoperability requirements for blockchain systems. Furthermore, we contribute to the understanding of blockchain interoperability beyond the technical and semantic perspectives of the extant literature and highlighting the legal and organizational requirements for enabling enterprises blockchain interoperability.

The present study has some limitations. Only 5 databases and a general google search were used to source articles. Selecting specific databases over others leads to selection bias, and may result in the exclusion of other articles that may have been relevant. Furthermore, studies that were not in English were excluded even though they may have been applicable. Moreover, the findings of the review are not exhaustive. This is due to the scarcity of literature on blockchain interoperability requirements. Therefore, a future research agenda could focus on exploring the legal and organizational requirements in more detail. Additionally, future studies can explore the domain specific requirements in various industries. In addition, the requirements for connecting different blockchains such as, permissioned to permissionless, permissioned to other permissioned as well as connecting blockchains to non-blockchains still need further investigation. Therefore, future research could focus on investigating the requirements from these perspectives. It would also benefit the literature to define requirements with cross-chain integration mechanisms as these are often suggested as solutions for enabling blockchain interoperability.

References

1. Min, H.: Blockchain technology for enhancing supply chain resilience. Bus. Horiz. **62**(1), 35–45 (2019)
2. Zhang, P., D.C. Schmidt, J. White, G. Lenz, Chapter One - Blockchain Technology Use Cases in Healthcare, in Advances in Computers, P. Raj and G.C. Deka, Editors. 2018, Elsevier. p. 1–41
3. Polyviou, A., Velanas, P., Soldatos, J.: Blockchain technology: financial sector applications beyond cryptocurrencies. Multidiscip. Digital Publish. Inst. Proceedings **28**(1), 7 (2019)
4. Morkunas, V.J., Paschen, J., Boon, E.: How blockchain technologies impact your business model. Bus. Horiz. **62**(3), 295–306 (2019)
5. Nodehi, T., Zutshi, A., Grilo, A., Rizvanovic, B.: EBDF: The enterprise blockchain design framework and its application to an e-Procurement ecosystem. Comput. Ind. Eng. **171**, 108360 (2022)
6. IEEE. 2022 IEEE International Conference on Blockchain and Cryptocurrency Cross-chain Workshop, ICBC-CROSS 2022.In: IEEE International Conference on Blockchain and Cryptocurrency Crosschain Workshop, ICBC-CROSS 2022 (2022)
7. European Commission, European Interoperability Framework (EIF). 2017, European Commission
8. Pang, Y.: A new consensus protocol for blockchain interoperability architecture. IEEE Access **8**, 153719–153730 (2020)
9. Pillai, B., Biswas, K., Hou, Z., Muthukkumarasamy, V.: Cross-Blockchain technology: integration framework and security assumptions. IEEE Access **10**, 41239–41259 (2022)
10. Sigwart, M., Frauenthaler, P., Spanring, C., Sober, M., Schulte, S.: Decentralized Cross-Blockchain Asset Transfers. In: 2021 3rd International Conference on Blockchain Computing and Applications, BCCA 2021 (2021)

11. Belchior, R., Vasconcelos, A., Guerreiro, S., Correia, M.: A survey on blockchain interoperability: past, present, and future trends. ACM Comput. Surv., **54**(8), p. Article 168 (2021)
12. Pillai, B., Hóu, Z., Biswas, K., Bui, V., Muthukkumarasamy, V.: Blockchain interoperability: performance and security trade-offs. In: Proceedings of the 20th ACM Conference on Embedded Networked Sensor Systems. 2023, Association for Computing Machinery: Boston, Massachusetts, pp. 1196–1201 (2023)
13. Ndlovu, K., Mars, M., Scott, R.E.: Interoperability frameworks linking mHealth applications to electronic record systems. BMC Health Serv. Res. **21**(1), 459 (2021)
14. Zhou, Q., Lee, Y.-S.: Blockchain interoperability mechanism. J. Korea Inst. Inform. Commun. Eng. **25**(11), 1676–1686 (2021)
15. Koens, T., Poll, E.: Assessing interoperability solutions for distributed ledgers. Pervasive Mob. Comput. **59**, 101079 (2019)
16. Wang, G., Wang, Q., Chen, S.: Exploring Blockchains Interoperability: A Systematic Survey. ACM Comput. Surv. **55** (2023)
17. Qasse, I.A., Talib, M.A., Nasir, Q.: Inter blockchain communication: a survey. In: Proceedings of the ArabWIC 6th Annual International Conference Research Track. 2019, Association for Computing Machinery: Rabat, Morocco. p. Article 2 (2019)
18. Monika, Bhatia, R.: Interoperability solutions for blockchain. In: 2020 International Conference on Smart Technologies in Computing, Electrical and Electronics (ICSTCEE). (2020)
19. Belchior, R., Riley, L., Hardjono, T., Vasconcelos, A., Correia, M.: Do you need a distributed ledger technology interoperability solution? Distrib. Ledger Technol. (2022)
20. Kitchenham, B.: Procedures for performing systematic reviews. Keele, UK, Keele University **2004**(33), 1–26 (2004)
21. Garousi, V., Felderer, M., Mäntylä, M.V.: Guidelines for including grey literature and conducting multivocal literature reviews in software engineering. Inf. Softw. Technol. **106**, 101–121 (2019)
22. Bradach, B., Nogueira, J., Llambías, G., González, L., Ruggia, R.: A gateway-based interoperability solution for permissioned blockchains. In: 2022 XVLIII Latin American Computer Conference (CLEI) (2022)
23. Jin, H., Dai, X., Xiao, J.: Towards a novel architecture for enabling interoperability amongst multiple blockchains. In: 2018 IEEE 38th International Conference on Distributed Computing Systems (ICDCS) (2018)
24. Hardjono, T., Lipton, A., Pentland, A.: Toward an interoperability architecture for blockchain autonomous systems. IEEE Trans. Eng. Manage. **67**(4), 1298–1309 (2020)
25. Zhang, P., White, J., Schmidt, D.C., Lenz, G.: Applying software patterns to address interoperability in blockchain-based healthcare apps. arXiv preprint arXiv:1706.03700, (2017)
26. World Bank Group, Blockchain Interoperability (2020)
27. Sonkamble, R.G., Phansalkar, S.P., Potdar, V.M., Bongale, A.M.: Survey of interoperability in electronic health records management and proposed blockchain based framework: MyBlockEHR. IEEE Access **9**, 158367–158401 (2021)
28. Ghosh, B.C., et al.: Decentralized cross-network identity management for blockchain interoperation. In: 2021 IEEE International Conference on Blockchain and Cryptocurrency (ICBC) (2021)
29. Liu, S., Mu, T., Xu, S., He, G.: Research on cross-chain method based on distributed Digital Identity. In: The 2022 4th International Conference on Blockchain Technology, pp. 59–73. Association for Computing Machinery, Shanghai, China (2022)
30. Al-Rakhami, M., Al-Mashari, M.: Interoperability approaches of blockchain technology for supply chain systems. Business Process Management Journal (2022) (ahead-of-print)

31. Hewett, N., van Gogh, M., Pawczuk, L.: Inclusive Deployment of Blockchain for Supply Chains: Part 6 – A Framework for Blockchain Interoperability. 2020, World Economic Forum: Switzerland
32. Lipton, A. Hardjono, T.: Blockchain Intra- and Interoperability, in Springer Series in Supply Chain Management. 2022. pp. 1–30 (2022)
33. Pillai, B., Biswas, K., Hou, Z., Muthukkumarasamy, V.: Level of conceptual interoperability model for blockchain based systems. In: IEEE International Conference on Blockchain and Cryptocurrency Crosschain Workshop, ICBC-CROSS 2022 (2022)
34. Sober, M., Sigwart, M., Frauenthaler, P., Spanring, Kobelt, M., Schulte, S.: Decentralized cross-blockchain asset transfers with transfer confirmation. Cluster Comput. J. Netw. Softw. Tools Appl. (2022)
35. Robinson, P.: Survey of crosschain communications protocols. Comput. Netw. **200**, 108488 (2021)
36. Madine, M., Salah, K., Jayaraman, R., Al-Hammadi, Y., Arshad, J., Yaqoob, I.: AppxChain: application-level interoperability for blockchain networks. IEEE Access **9**, 87777–87791 (2021)
37. Lemrabet, Y., Bigand, M., Clin, D., Benkeltoum, D., Bourey, J.-P.: Model driven interoperability in practice: preliminary evidences and issues from an industrial project. In: Proceedings of the First International Workshop on Model-Driven Interoperability. Association for Computing Machinery: Oslo, Norway. pp. 3–9 (2010)
38. Reegu, F.A., et al.: Systematic assessment of the interoperability requirements and challenges of secure blockchain-based electronic health records. Security and Communication Networks (2022)
39. García-Bañuelos, L., Ponomarev, A., Dumas, M., Weber, I.: Optimized execution of business processes on blockchain. In: Business Process Management: 15th International Conference, BPM 2017, Barcelona, Spain, September 10–15, 2017, Proceedings 15 Springer (2017)
40. Viriyasitavat, W., Bi, Z., Hoonsopon, D.: Blockchain technologies for interoperation of business processes in smart supply chains. J. Ind. Inf. Integr. **26**, 100326 (2022)
41. Lehmann, M.: Who owns bitcoin: private law facing the blockchain. Minn. JL Sci. & Tech. **21**, 93 (2019)
42. European Commission, Study on Blockchains: Legal, governance and interoperability aspects (2020)

Deep Learning and Healthcare Applications

PENN: Phase Estimation Neural Network on Gene Expression Data

Aram Ansary Ogholbake and Qiang Cheng(✉)

University of Kentucky, Lexington, KY 40526, USA
{aram.ansary,qiang.cheng}@uky.edu

Abstract. With the continuous expansion of available transcriptomic data like gene expression, deep learning techniques are becoming more and more valuable in analyzing and interpreting them. The National Center for Biotechnology Information Gene Expression Omnibus (GEO) encompasses approximately 5 million gene expression datasets from animal and human subjects. Unfortunately, the majority of them do not have a recorded timestamps, hindering the exploration of the behavior and patterns of circadian genes. Therefore, predicting the phases of these unordered gene expression measurements can help understand the behavior of the circadian genes, thus providing valuable insights into the physiology, behaviors, and diseases of humans and animals. In this paper, we propose a novel approach to predict the phases of the un-timed samples based on a deep neural network architecture. It incorporates the potential periodic oscillation information of the cyclic genes into the objective function to regulate the phase estimation. To validate our method, we use mouse heart, mouse liver and temporal cortex of human brain dataset. Through our experiments, we demonstrate the effectiveness of our proposed method in predicting phases and uncovering rhythmic pattern in circadian genes.

1 Introduction

Circadian rhythms, which follow roughly a 24-h cycle, have significant impacts on health and behavior of humans and animals. For instance, body temperature, blood pressure and heart rate follow circadian rhythms. The underlying reason behind these rhythms is primarily driven by the clock genes. The core clock genes in mammals include Bmal1 (brain and muscle aryl-hydrocarbon receptor nuclear translocator-like 1), CLOCK (Circadian Locomotor Output Cycles Kaput), Cry1,2 (Cryptochrome1,2) and Per1,2,3 (Period1,2,3) [19]. These clock genes can be affected by some external factors such as irregular dark-light cycles, environmental stresses and jetlag. Disrupted clock gene expression could lead to physical problems such as cancer, obesity, sleep disorder, and diabetes [8]. Moreover, disrupted circadian rhythms could aggravate symptoms in individuals with mental health issues such as major depressive disorder (MDD), bipolar disorder (BD), anxiety, and schizophrenia (SZ) [18]. Clock gene dysfunction is also linked with neurodegenerative diseases, e.g., Alzheimer's disease [13,17]. Therefore, it

M. Younas et al. (Eds.): Deep-BDB 2023, LNNS 768, pp. 59–67, 2023.
https://doi.org/10.1007/978-3-031-42317-8_5

is important to study the gene circadian rhythms for understanding the mechanisms of physiology and diseases.

Currently, availability of sample times in gene expression data is key to identifying which genes are rhythmic. However, majority or even most of the biological samples and thus gene expression samples do not have timestamps, even though there are millions of such samples in the National Center for Biotechnology Information Gene Expression Omnibus (GEO). Therefore, an important task in circadian study is to computationally estimate the time labels of the samples. Moreover, due to invasiveness, it is hard to obtain in vivo biological samples and, in particular, human brain samples usually do not have available timestamps. Human subjects are vulnerable to the potential health risks associated with collecting data at special time intervals over multiple days. While gene expression data collected from postmortem brain samples may have information about time of death (TOD), such time usually is inaccurate. Hence, machine learning-based methods which can estimate sample time, or phases, on un-timed dataset in an unsupervised manner, will be particularly useful.

There have been several algorithms to order periodic data using machine learning techniques, e.g., [2–4,7,12], which will be described in more detail in Sect. 2. Among them, CYCLOPS [2] is a popular method based on unsupervised learning. The phase of each sample is predicted using a circular auto-encoder architecture, which non-linearly projects each high-dimensional sample onto a unit circle. In our paper, we propose Phase Estimation Neural Network on gene expression data (PENN) method which is inspired by this work and it aims to predict the phase of each sample within unordered gene expression data. PENN, also uses circular auto-encoder where the bottleneck layer outputs are used to calculate the phase for each sample. The objective function of CYCLOPS is the mean squared error to reconstruct the input. However, it does not consider the geometry of gene oscillations that can offer useful regularization to the nonlinear projection. However, in PENN, we introduce a novel objective function which is used to jointly fit (eigen)genes to cosine curves and reconstruct the original input. Experimental results on cross-species/tissue data validate the effectiveness of our proposed approach.

2 Related Work

While having important use cases, the methods in the state of the art also demonstrate limitations in important applications. For instance, Zeitzeiger [7] and TimeSignature [4] need annotated circadian times available for supervised learning. However, most of the gene expression data do not have recorded timestamps which limits the usage of supervised learning methods in prediction of phases. Zeitzeiger [7] is a method which uses supervised learning to estimate the circadian time from a high-dimensional observation. Training data consists of a corresponding time for each sample and a matrix of samples and features. Then, a new matrix is constructed where the time-dependant densities of features are discretized and scaled. Then sparse principal components (SPCs) of the new matrix is calculated and used to project the training data into SPC space. And finally the network uses a maximum likelihood to make predictions. In TimeSignature [4], firstly gene expression data is rescaled and normalized. Then, the normalized data is fitted by a multivatiate regression model with elastic net regularization. Subsequently, the fitted model is used to predict time of day of new samples. PENN which is an unsupervised learning method, does not depend on any annotated data in order to predict the phase of each sample.

There are also methods focusing on single-cell RNA-sequence (scRNA-seq) data. For example, Oscope [12] is for cell cycle estimation rather than circadian phase estimation, which is highly sensitive to inter-subject variations. This algorithm first fits a two-dimensional sinusoidal function to all gene pairs with similar frequency and chooses the best fit among them. Then it clusters these gene into groups using K-medoids algorithm. And at last, it recovers the cell order in each gene group. Tempo [3] uses a Bayesian approach to maximize the posterior probability of individual cells' phases over scRNA-seq data rather than the tissue samples' phases.

Notably, the cell cycle and the circadian rhythm are two different biological processes and their estimations are also different [3,11,16]. For example, the former leads to the duplication and division of a cell, consisting of four phases, G1, S, G2, and M; in contrast, the latter regulates various physiological and behavior processes, consisting of two phases, active and rest. Also, the cell cycle has a variable duration dependent on the cell type and external factors, e.g., growth factors, nutrients, and DNA damage; on the other hand, circadian rhythm has a fixed period of about 24 h entrained by external cues, e.g., light and temperature. These two different processes are linked with variable genes: Hundreds of core genes are known to oscillate over the cell cycle, while only about 20 core circadian clock genes are known for the circadian rhythm. Also, many core cell cycle genes are highly expressed, while core circadian clock genes are moderately expressed. Further, the circadian controlled genes are driven by the core clock or other transcription factors coupled with the core clock; they are cell-type specific over the circadian cycle and their identifies are unknown ahead of time. On the other hand, the genes involved in the cell cycle are relatively independent of cell type and are known ahead of time. These differences in the processes and involved genes lead to different ways for estimation of circadian rhythmic phases and cell cycles when based on untimed gene expression data. In this paper, we will focus on circadian phase estimation on untimed transcriptomic data at the tissue level (e.g., bulk RNA-seq) rather than scRNA seq data.

3 Method

The proposed PENN model addresses the task of unordered samples' phase prediction over transcriptomic data, in particular, gene expression. The original data typically include thousands or tens of thousands of genes, among which only a proportion are cycling genes. To reduce the dimensionality of the data, similar to CYCLOPS [2], PENN first uses Principal Component Analysis (PCA) on the dataset and calculates the principal components (called eigengenes). A proper number of top eigengenes typically contain sufficient information to determine the relative order of samples and they are also cycling with a phase of about 24 h [1,2]. Because of the datasets' variable sample sizes, PENN will leverage a neural network architecture that is applicable to small sample sizes. Auto-encoder (AE) architecture is a suitable candidate. By using reconstruction of the input data as an objective, it uses the data itself to provide supervision rather than depend on any annotated labels in training the network, thus robust to small sample size. Therefore, similar to CYCLOPS, the AE architecture is adopted in PENN to predict the phase of each sample. The dimension-reduced data with PCA will be fed into AE in which the bottleneck layer consists of a circular node [10]. Specifically, the two coupled neurons, u_i and v_i for the i-th sample, $i = 1, \cdots, m$, are restricted to be on a unit circle, whose phase ϕ_i is computed as

$$\phi_i = \arctan(\frac{u_i}{v_i}). \tag{1}$$

In CYCLOPS, the objective of the loss function is to minimize the discrepancy between the reconstructed output and the original input, without incorporating any gene information. The high-dimensional samples are typically located on or around a sinusoidal curve and the expression values of eigengenes or cycling genes also reside on or around sinusoidal curves. This rich structural information for samples and cycling genes, however, is not utilized in CYCLOPS. In this paper, we propose an innovative objective function to incorporate gene structural information into the optimization process. By incorporating this gene-specific information to regularize the phase estimation, our objective function improves the overall performance of the model.

3.1 Objective Function of PENN

Let's denote samples by $x_1, x_2, ..., x_m$ where m is the number of samples. Each sample x_i consists of n features (gene values) where n is the number of eigengenes. Let's denote eigengenes by $g_1, g_2, ..., g_n$. As we mentioned, the bottleneck layer is used to estimate the phase of each sample x_i. If all samples are estimated correctly, then each eigengene can be fitted by a cosine curve:

$$\hat{Y}_i^{(g)} = L^{(g)} + A^{(g)} \cos(\omega^{(g)} \hat{\phi}_i + \phi^{(g)}) \tag{2}$$

where $L^{(g)}$ is the average level of the g-th eigengene, $A^{(g)}$ is the amplitude for the g-th eigengene, $\phi^{(g)}$ is used for the phase shift and $\frac{2\pi}{\omega}$ is the period. We use this in our objective function, so the network estimates the best phase for each sample that leads into the best cosine curve fit to each gene. Our objective function is then defined as:

$$\mathfrak{L} = \frac{\lambda}{m} \sum_{i=1}^{m} \frac{1}{n} \sum_{g=1}^{n} [x_i^{(g)} - (L^{(g)} + A^{(g)} \cos(\omega^{(g)} \hat{\phi}_i + \phi^{(g)}))]^2 + \frac{1}{m} \sum_{i=1}^{m} \|x_i - \hat{x}_i\|^2 \tag{3}$$

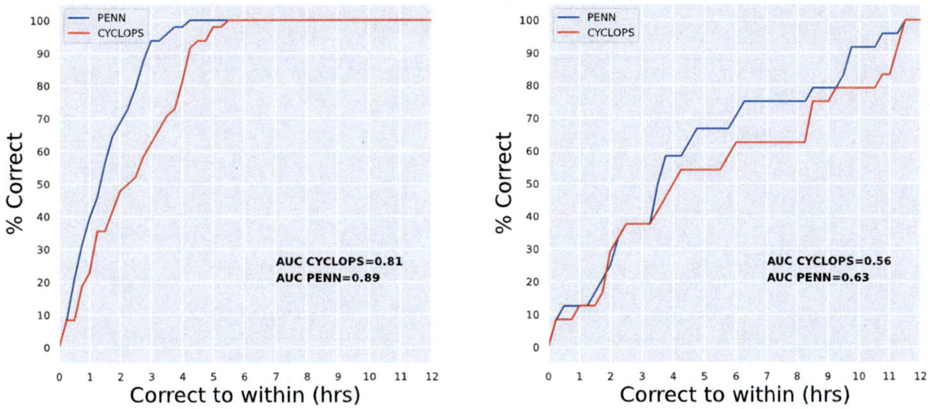

Fig. 1. Accuracy of CYCLOPS compared to PENN on mouse liver data(GSE11923) (left) and Mouse Heart (GSE54650) (right). The y-axis shows the fraction of correctly predicted samples, and the x-axis shows the size of being correct to within hours.

where $\hat{x_i}$ is the output of the auto-encoder and λ is a balancing factor. In this formula, $L^{(g)}$, $A^{(g)}$, $\phi^{(g)}$ and $\omega^{(g)}$ are learnable variables given some initial values and they will be trained in the network. Based on Eq. 3, our loss function has two parts. The first part of the equation is the error of fitting observations with the cosine curve using estimated phases. The second part of the loss function is the auto-encoder reconstruction loss.

4 Results

4.1 Dataset

In order to validate our proposed method, we used three different dataset: mouse liver, mouse heart, and temporal cortex of human brain. The mouse liver dataset (GSE11923) consists of 48 samples obtained from the livers of 3–5 mice at each time point, with samples taken every hour [6]. The mouse heart dataset is from a bigger data (GSE54650) which contains 288 samples covering 12 tissues. We focused on the heart tissue which was sampled every 2 h for 48 h [20]. The temporal cortex dataset contains 71 subjects (35 females and 35 males). However, there was no timestamp or TOD recorded for each sample in this dataset [9,14]; thus, we will verify our algorithm by confirming that it can properly uncover the rhythmicity for many existing clock genes well known in the literature.

4.2 Experiments

We conducted two different sets of experiments: one for annotated data on which we have access to timestamps and the other for unannotated data.

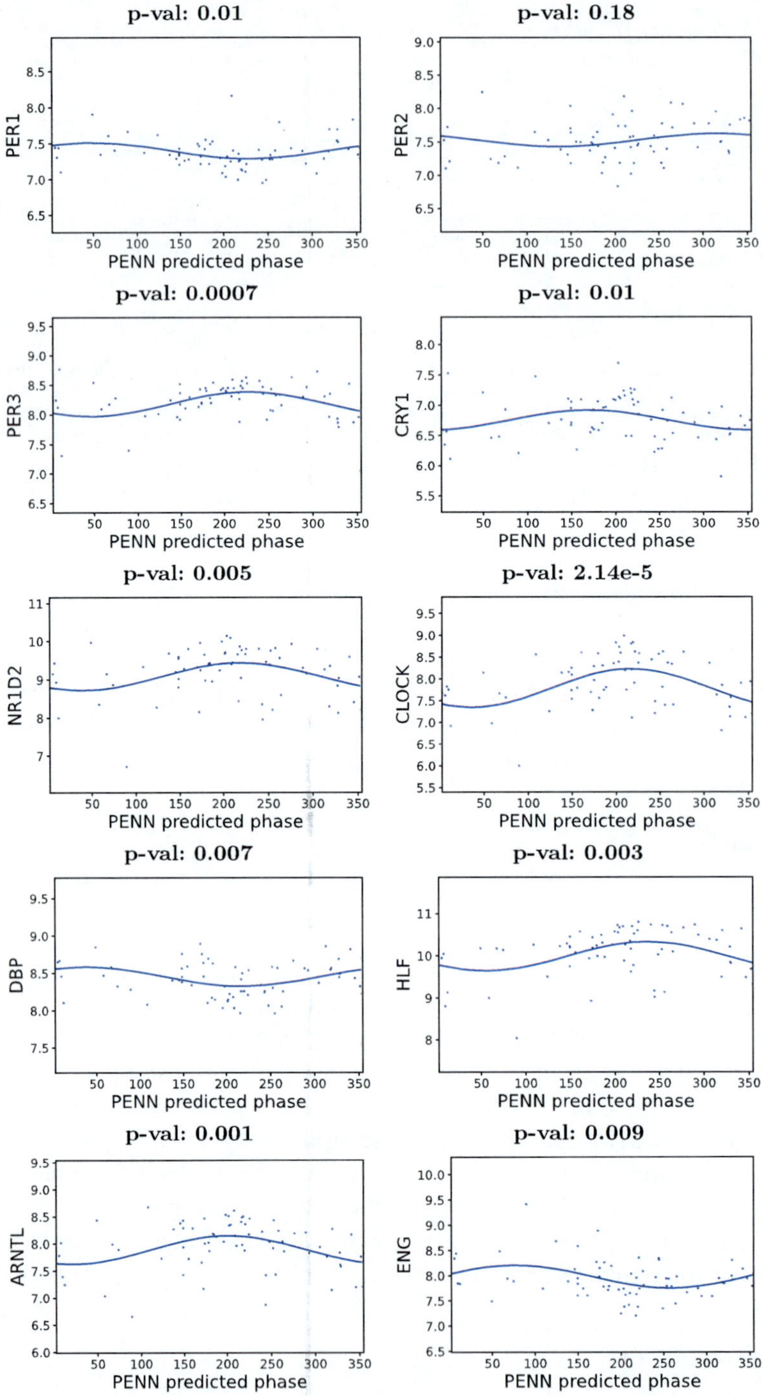

Fig. 2. Plots of gene expression versus predicted phase for 10 circadian genes on the temporal cortex tissue of human brains (GSE131617). P-values of the sinusoidal curve fitting are shown above each plot.

For the annotated data (GSE11923 and GSE54650), we compared PENN with CYCLOPS. We plotted the fraction of correctly predicted samples versus the size of being correct to within hours [4]. The perfect prediction has a normalized area under the curve (AUC) of 1. As shown in Fig. 1, PENN improved upon CYCLOPS in AUC. On the mouse liver dataset, PENN showed an AUC of 0.89, which was 8% higher than CYCLOPS. The AUC on the mouse heart dataset also improved upon CYCLOPS from 0.56 to 0.63.

For the temporal cortex of human brain dataset, we plotted the gene expression values vs the predicted phase for some of the circadian genes (Fig. 2). These genes are known clock genes in the literature. We used CosinorPy [15] to fit cosine curve for each gene if there is any circadian rhythmic pattern. We observe that all of the genes show circadian rhythmic patterns and 9 out of 10 are fitted with p-values less than 0.05. The results demonstrate that our method can successfully recover circadian phases, which can be subsequently used to uncover rhythmic patterns of known circadian genes [5].

4.3 Implementation

We normalized our data using z-score normalization:

$$x_{ij}^{1} = \frac{x_{ij}^{0} - \mu}{\sigma} \tag{4}$$

where x_{ij}^{0} is original data value at sample i and gene j and x_{ij}^{1} is the normalized data value. μ is the average over all samples and genes. σ is the standard deviation over all samples and genes.

We trained our model using Keras. We used auto-encoder architecture to predict the phases of each sample through the bottleneck layer. The encoder and decoder parts each consisted of two layers, where each layer was fully connected (dense). The activation function for all layers other than the bottleneck layer was linear, which means the values only pass through these layers. The values of the coupled neurons in the bottleneck layer after activation were normalized to satisfy the restriction on being on a unit circle. The phase was then calculated using these coupled neurons. We also defined four trainable variables L, A, ϕ, and ω which were tensors of shape $(1, n)$, where n was the number of genes. These variables will be learned based on our defined objective function in Sect. 3.1. The Glorot Uniform initializer was used on each layer, which sampled the initial weights from a uniform distribution. We used SGD as the optimizer and trained our model for 1000 epochs with a learning rate of 0.5.

5 Conclusion

We proposed PENN, a novel approach for the prediction of unordered gene expression data. This deep learning based method incorporates gene structural information while optimizing the objective function. We use a circular auto-encoder based architecture. Our proposed objective function consists of two parts: fit loss and reconstruction loss. The fit loss is the error of fitting observations with the cosine curve using predicted phases. For reconstruction loss, we use the mean squared error between the original input and the reconstructed output. We validated our proposed method on both annotated and unlabeled datasets. Our results show improved accuracy compared to the state-of-the-art method. Moreover, it shows circadian rhythmic patterns on circadian genes with our predicted times on samples. In our future work, we intend to expand our method by incorporating an additional human dataset that contains a larger number of data points. These datasets are expected to be more intricate, requiring a more sophisticated neural network to effectively handle the increased complexity.

Acknowledgement. This study was partially supported by NIH R21 AG070909-01, P30 AG072946-01, and R01 HD101508-01.

References

1. Alter, O., Brown, P.O., Botstein, D.: Singular value decomposition for genome-wide expression data processing and modeling. Proc. Natl. Acad. Sci. **97**(18), 10101–10106 (2000)
2. Anafi, R.C., Francey, L.J., Hogenesch, J.B., Kim, J.: Cyclops reveals human transcriptional rhythms in health and disease. Proc. Natl. Acad. Sci. **114**(20), 5312–5317 (2017)
3. Auerbach, B.J., FitzGerald, G.A., Li, M.: Tempo: an unsupervised bayesian algorithm for circadian phase inference in single-cell transcriptomics. Nature Commun. **13**(1), 6580 (2022)
4. Braun, R., et al.: Universal method for robust detection of circadian state from gene expression. Proc. Natl. Acad. Sci. **115**(39), E9247–E9256 (2018)
5. Chen, C.Y., et al.: Effects of aging on circadian patterns of gene expression in the human prefrontal cortex. Proc. Natl. Acad. Sci. **113**(1), 206–211 (2016)
6. Hughes, M.E., et al.: Harmonics of circadian gene transcription in mammals. PLoS Genet. **5**(4), e1000442 (2009)
7. Hughey, J.J., Hastie, T., Butte, A.J.: Zeitzeiger: supervised learning for high-dimensional data from an oscillatory system. Nucleic acids research **44**(8), e80–e80 (2016)
8. Khan, S.: Health risks associated with genetic alterations in internal clock system by external factors. Int. J. Biol. Sci. **14**(7), 791–798 (2018)
9. Kikuchi, M., et al.: Disruption of a rac1-centred network is associated with alzheimer's disease pathology and causes age-dependent neurodegeneration. Human Molec. Genet. **29**(5), 817–833 (2020)
10. Kirby, M.J., Miranda, R.: Circular nodes in neural networks. Neural Comput. **8**(2), 390–402 (1996)
11. Kowalczyk, M.S., et al.: Single-cell rna-seq reveals changes in cell cycle and differentiation programs upon aging of hematopoietic stem cells. Genome Res. **25**(12), 1860–1872 (2015)
12. Leng, N., et al.: Oscope identifies oscillatory genes in unsynchronized single-cell rna-seq experiments. Nat. Methods **12**(10), 947–950 (2015)
13. Li, P., et al.: Circadian disturbances in alzheimer's disease progression: a prospective observational cohort study of community-based older adults. Lancet Healthy Longevity **1**(3), e96–e105 (2020)

14. Miyashita, A.: Genes associated with the progression of neurofibrillary tangles in alzheimer's disease. Transl. Psychiatry **4**(6), e396–e396 (2014)
15. Moškon, M.: Cosinorpy: a python package for cosinor-based rhythmometry. BMC Bioinf. **21**, 1–12 (2020)
16. Santos, A., Wernersson, R., Jensen, L.J.: Cyclebase 3.0: a multi-organism database on cell-cycle regulation and phenotypes. Nucleic Acids Res. **43**(D1), D1140–D1144 (2015)
17. Thome, J., Coogan, A.N., Woods, A.G., Darie, C.G., Häßler, F.: Clock genes and circadian rhythmicity in alzheimer disease. J. Aging Res. **2011** (2011)
18. Walker, W.H., Walton, J.C., DeVries, A.C., Nelson, R.J.: Circadian rhythm disruption and mental health. Transl. Psychiatry **10**(1), 28 (2020)
19. Zhang, J., Sun, R., Jiang, T., Yang, G., Chen, L.: Circadian blood pressure rhythm in cardiovascular and renal health and disease. Biomolecules **11**(6), 868 (2021)
20. Zhang, R., Lahens, N.F., Ballance, H.I., Hughes, M.E., Hogenesch, J.B.: A circadian gene expression atlas in mammals: implications for biology and medicine. Proc. Natl. Acad. Sci. **111**(45), 16219–16224 (2014)

MRIAD: A Pre-clinical Prevalence Study on Alzheimer's Disease Prediction Through Machine Learning Classifiers

Jannatul Loba[1], Md. Rajib Mia[1](✉), Imran Mahmud[1], Md. Julkar Nayeen Mahi[1](✉), Md. Whaiduzzaman[2], and Kawsar Ahmed[3]

[1] Department of Software Engineering, Daffodil International University, Dhaka, Bangladesh
rajib.swe@diu.edu.bd, mahi.1992@gmail.com
[2] School of Information Systems, Queensland University of Technology, Brisbane, Australia
[3] Department of Electrical and Computer Engineering, University of Saskatchewan, Saskatoon, Canada

Abstract. Alzheimer's disease (AD) is a neurological illness that worsens with time. The aged population has expanded in recent years, as has the prevalence of geriatric illnesses. There is no cure, but early detection and proper treatment allow sufferers to live normal lives. Furthermore, people with this disease's immune systems steadily degenerate, resulting in a wide range of severe disorders. Neuroimaging Data from magnetic resonance imaging (MRI) is utilized to identify and detect the disease as early as possible. The data is derived from the Alzheimer's Disease Neuroimaging Initiative (ADNI) collection of 266 people with 177 structural brain MRI imaging, DTI, and PET data for intermediate disease diagnosis. When neuropsychological and cognitive data are integrated, the study found that ML can aid in the identification of preclinical Alzheimer's disease. Our primary objective is to develop a model that is reliable, simple, and rapid for diagnosing preclinical Alzheimer's disease. According to our findings (MRIAD), the Logistic Regression (LR) model has the best accuracy and classification prediction of about 98%. The ML model is also developed in the paper. This article profoundly, describes the possibility to getting into Alzheimer's disease (AD) information from the pre-clinical or non-preclinical trial datasets using Machine Learning Classifier (ML) approaches.

Keywords: Alzheimer's Disease · Machine Learning · MRI · Logistic Regression

1 Introduction

Over 50 million individuals worldwide have dementia, and that figure is anticipated to climb to 152 million by 2050 [1] if 12.7% of the world's population is

M. Younas et al. (Eds.): Deep-BDB 2023, LNNS 768, pp. 68–80, 2023.
https://doi.org/10.1007/978-3-031-42317-8_6

younger than 60 years old. Alzheimer's disease is one of several kinds of dementia. It results in irreparable brain damage, cognitive loss, and death [2]. Affects memory, cognition, and behavior [3] and makes speaking, reading, and writing difficult. There are three distinct phases of Alzheimer's disease: extremely mild, mild, and moderate [4]. Changes in the brain, blood, and CSF occur before behavioral indications in the early stages of Alzheimer's disease. Second, moderate cognitive impairment (MCI), memory problems, and other cognitive abnormalities began to proliferate. Finally, memory and behavioral issues interfere with daily living [5]. Dehydration, malnutrition, and infections are common in the latter stages of the illness and can lead to death. There is no known treatment for Alzheimer's. After 3 to 10 years, the patient dies. Early detection is critical for the successful treatment of Alzheimer's disease or perhaps slowing its progression. Alzheimer's disease is related to metabolic abnormalities in the brain, structural atrophy, and pathological amyloid deposits [6]. Amyloid protein promotes cell death and impedes signal transmission, resulting in memory loss, difficulty thinking, focusing, making decisions, and cognitive impairment [7]. Mini-Mental State Examination (MMSE), Alzheimer's disease Assessment Scale (ADAS), Montreal Cognitive Assessment (MoCA), Positron emission tomography (PET), Magnetic resonance imaging (MRI), Diffusion tensor imaging (DTI), Clinical dementia rating scale (CDR) and a detailed history of patients; all required for a proper medical assessment to predict AD progression [8]. Standard scale psychological examinations include the MMSE, MoCA, and ADAS; DTI and PET examine the location, direction, and anisotropy of white matter pathways, as well as regional glucose metabolism and amyloid deposition. MRI detects and quantifies neurodegenerative alterations (cortical thinning, brain shrinkage, and regional tissue density) [9]. The development of mild memory loss to vascular dementia is illustrated and scored using MRI of brain areas such as the hippocampus, parahippocampus, and amygdalae [10,11]. MRI creates a three-dimensional picture of the internal structure of the brain. In this article, LR, KNN, and GNB, the most prominent classification algorithms, are utilized to predict the progression of preclinical and non-preclinical Alzheimer's disease. The Classifiers described above were trained and assessed independently utilizing level-based train-test split data scores. SelectkBest chose the following 22 characteristics after deleting all DTI data: Gender01, Marry123, ABETA1700, ABETA12, TAU80, PTAU8, AD123, AGE, EDU, PTRACCAT, APOE4. The highest accuracy of Logistic Regression is 98%. Our overall technique correctly forecasts and reviews the effective result of the preclinical stage of Alzheimer's disease patients.

Table 1 following the previous work overview, the most frequent limitations in the majority of papers are datasets and classification models accuracy. Many research work classification model accuracies are quite satisfying. After using a dataset containing several neuroimaging and cognition data and models are predicted to have higher accuracy in this study.

Table 1. Model comparison with other previous studies

Reference	Target	Best Classifier	Accuracy
Rallabandi. et al. 2020 [32]	CN, EMCI, LMCI	SVM	75%
Neelaveni, J. et al. 2020 [8]	AD	SVM	85%
Alickovic, E. et al. 2020 [21]	AD	RF	85.77%
Ghoraani, B. et al. 2020 [33]	MCI, AD	SVM	86%
Shah, A. et al. 2020 [23]	Early detection of AD	Voting Classifier Algorithm	86%
EKE, C. S. et al. 2020 [1]	AD	SVM	89%
Marzban, E. N. et al. 2020 [15]	MCI, AD	CNN	MD: 94% GM:84%
Yamashita, A. Y. et al. 2021 [9]	AD	SVM	95.1%
Alshammari, M. et al. 2021 [13]	AD	CNN	97%
Proposed Study	**Preclinical AD, Non- preclinical AD**	**Logistic Regression**	**98%**

2 Related Work

Many studies have found ways to diagnose AD using various methods. A conventional SVM and a CNN based on structural MRI image data were used to predict AD and CN [12]. In this article, CNN is used to detect Alzheimer's disease on MRI scans [13]. The suggested approach of instance-based feature mapping detected MCI using multiple-kernel learning [14]. The CNN program was used to predict AD using MRI and DTI image data [15]. The AlexNet model retrieved detailed characteristics from CNN and used SVM, KNN, and RF to suggest an AD stage detection [16]. Another study employed structural brain MRI imaging data to train the LR model, which compares and predicts Alzheimer's disease categorization [17]. Likewise, this work used the SVM and KNN algorithms using MRI data to predict AD characteristics [18]. A research used SVM models to diagnose AD based on voice signals translated into Syntactic, Semantic, and Pragmatic values [19], although merely investigating audio voice data revealed AD is rather difficult. Another research, Automated Detection Based on Histogram and Random Forest [20], is insufficient to diagnose AD since neither of these techniques can identify AD independently. Participants in a research were provided MRI, PET, and numerous biological markers such as clinical and cognitive evaluation data, and SVM models were used to differentiate between MCI and AD [21]. An study advocates employing Recurrent Neural Networks (RNNs) with Long Short-Term Memory (LSTM) and Gated Recurrent Unit (GRU) to identify memory impairment, such as CN, MCI, and AD [22]; however, training RNN takes a long time, and complex training processes are required. Another research uses Hard voting, Soft voting, DT, XGBoost, and RFC classifiers to eliminate all irrelevant values in 437 instances and achieve 86 percent performance [23]. To predict AD using RNN following CNN, the association between each subject's picture sequences was determined [24]. A CNN was utilized in several studies to predict AD, CN, and MCI utilizing diffusion maps and gray matter (GM) volumes [25].

Preclinical Alzheimer's disease develops before cognitive problems are diagnosed [26–28]. Early identification is crucial, and recent clinical studies [29–31] of disease-modifying drugs have shown that early therapy is best. The previous work did not adequately address preclinical and non-preclinical processes. Nonetheless, this study varies from previous investigations [32–36].This study employs cognitive and neuroimaging traits to properly determine whether an individual is preclinical or non-preclinical.

The remainder of the paper is arranged as follows. Section 2 discusses the data, data analysis, and modeling strategies used in the study. Section 2 outlines our techniques, and experimental design, and assesses and discusses the outcomes of a variety of machine-learning algorithms. Section 4 summarizes our findings and proposes recommendations for further research.

3 Research Methodology

Several machine-learning approaches and algorithms were developed in this work to identify neurodegenerative illnesses. Our new technique necessitates a series of process stages. Many computer vision applications make human life easier and safer. Figure 1 depicts each application briefly and visually describes all stages.

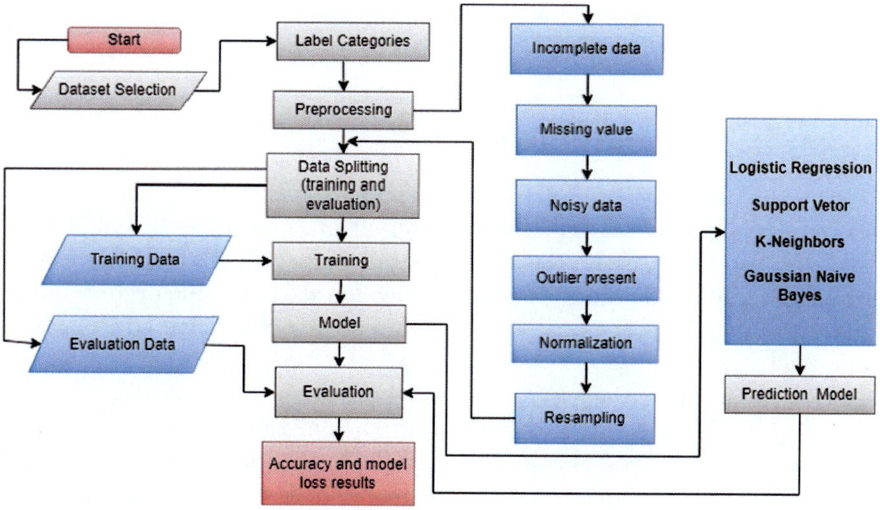

Fig. 1. Proposed Workflow Diagram

Figure 1 shows an overview of our research work in progress.

3.1 Development and Testing Approach

Machine learning is making computers smarter, and automation is rapidly expanding. It is concerned with how computers replicate or execute human learning behavior in order to acquire new information or abilities, reorganize existing knowledge structures, and continually improve their own performance. Use various ML algorithms to train for significant characteristics. For high accuracy, data is separated into two parts: training data (75% accuracy) and testing data (25% accuracy). The package representations are evaluated using four well-known performance parameters: accuracy (Ac), sensitivity (Sn), specificity (Sp), and F1-score (F1). Finally, using machine learning approaches to optimize simulation and experimentation, legitimate findings are shown differently by the Confusion matrix.

The LR model is the most typical application to solve classification problems. In this data binary and polynomial classes are the two different forms. The binary class was used for this data set because the output column has two prediction classes. For linearly separated data, logistic regression is appropriate. The risk of overfitting is low. The existing study is divided into two phases with independent and dependent patient data sets, such as stage 1 against stage 2 (stage 1's independent data set and stage 2's dependent data set).

$$log\frac{p(x)}{1-p(x)} = \beta_0 + \beta_1 X \tag{1}$$

In this equation, p is the estimated chance of the desired result, beta 0 is the intercept term, 1... I are the related regression coefficients with predictors X, and and i= 1, 2,... The i^n predictor has the number 177 [7]. This technique is easily expandable for class classification in this work when distinct data findings or differences can be isolated on a horizontal plane.

Equation environment, e.g.,

$$p(y) = \frac{1}{\sqrt{2\pi\sigma_y^2}} exp(-\frac{(x_i - \pi_y)^2}{2\sigma_y^2}) \tag{2}$$

In this procedure, y is a class, xi is an observation, $sigmay$ is the variance for the chosen class, and muy is the class's associated mean [18]. This research is scalable in predictors and data points, fast, and can generate real-time predictions of both discrete and continuous value; it is also simple to construct and implement in this dataset.

GNB is a simple categorization algorithm that works quickly. It is simple to work with little datasets. It is thought that the existence of one feature in a class is unconnected to the presence of other qualities. In the Navigator, all attributes are addressed independently. The recommended ensemble of classifiers may be trained using hypotheses for each class of sickness to forecast the likelihood and make predictions on new disease classes by selecting the class with the highest value of hypotheses. The primary benefit of a Gaussian classifier is that it requires very little characterization-relevant training data.

SVM is a directed study model that separates items using a hyperplane to classify them. It may be used for regression as well as classification. The margins are used to form hyperplanes. In two-dimensional (2D) space, this hyperplane is a line that splits into two parts, each on the opposite side [19]. SVM is being utilized in this work to determine an ideal hyperplane that gives a large minimal distance to the training data set. The process of selecting the closest training data points that fall within the features is known as partitioning. *equation* environment, e.g.,

$$d(x, y) = \sqrt{\sum_{i=1}^{n}(x_i - y_i)^2} \tag{3}$$

The distance between two points is calculated in this approach using the testing data point xi and the training data point yi, I = 1, 2, 3,.... n. The k value begins with 1, which represents the nearby size [20]. This research KNN makes real-time and short-term forecasts. The classifier here translates real-valued data into an output between 0 and 1, which is interpreted as the likelihood that the inputs belong to the positive class given the input values.

3.2 Data Source

In the medical business, the purpose of data processing techniques and machine learning is to aid in medical diagnosis. This article relied on the Alzheimer's Disease Neuroimaging Initiative (ADNI) database to collect and report data (adni.loni.usc.edu)[1].

The ADNI project seeks to investigate how serial MRI, PET, other biological markers, and clinical and neuropsychological examination might be used to detect the progression of MCI and AD. Each subject had cognitive and cerebral examinations, as well as physical diagnostics. The primary goal of this research is to analyze the utilization of magnetic fields and radio waves to monitor structural imaging of gray matter, white matter, and cerebrospinal fluid. Inside the sample, there are 177 people and 266 attributes. In all, 177 subjects were chosen for this study, including 107 men and 70 women between the ages of 55 and 90 from cognitively normal senior control participants. The label feature is represented by the dataset AV45AB12. Figure 2 depicts label attributes as well as all samples via count plot.

[1] The ADNI, a public-private partnership, was established in 2003 under the guidance of Principal Investigator Michael W. Weiner, MD. A comprehensive list of ADNI can be found at http://adni.loni.usc.edu/wpcontent/uploads/how/_to/_apply.

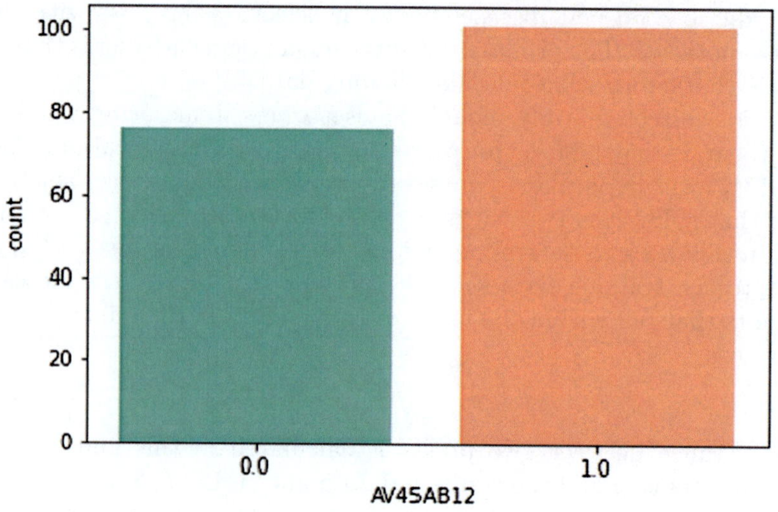

Fig. 2. Label feature count plot

Figure 2 shows the x-axis as label feature and the y-axis as label feature classes. There are two sorts of label values here: 0.0 for non-pre-clinical Alzheimer's disease (n = 76) and 1.0 for pre-clinical Alzheimer's disease (n = 101).

3.3 Data Preprocessing

This dataset contains a variety of data formats as well as unbalanced data. The missing data in this case is 999999. After converting 999999 to category NAN values. Label encoder is a tool for converting category data to numerical data. Numerical and missing data are transformed to median values, which are not impacted by outliers and may thus be estimated more accurately using ML approaches. Because training datasets are significantly unbalanced, the Synthetic Minority Oversampling Technique (SMOTE) data sampling technique is necessary to alleviate their class imbalance issue, which has no influence on median values. MinMaxScaler is used to balance or normalize data between 0 and 1.

3.4 Feature Selection

The major important characteristics will be employed during this feature selection approach, while worthless elements will be deleted. Because our dataset has several imbalances, select k-best function is effective for feature selection. It may reduce the time required while increasing accuracy [19].

4 Results and Discussion

The purpose of this study was to apply four supervised models to assess AD based on MRI data. LR, SVM, KNN, and GNB models must be identified. Using the dataset's noise, error, and geographical values to forecast real values. The performance of classifiers is measured using four distinct statistical metrics: (1) overall (CV) accuracy, (2) precision, (3) recall, and (4) F1-score.

$$precision = \frac{Truepositive}{Truepositive + Falsepositive} \tag{4}$$

$$Recall = \frac{Truepositive}{Truepositive + Falsenegative} \tag{5}$$

$$F1score = \frac{Precision * Recall}{Precision + Recall} \tag{6}$$

$$Accuracy = \frac{Truepositive + Truenegative}{Truepositive + Truenegative + Falsepositive + Falsenegative} \tag{7}$$

Precision and recall are two more helpful metrics. Out of the entire number of values, values that were appropriately categorized as positive are indeed positive. Accuracy is the proportion of correct predictions among all analyzed cases. F1 score combines model precision and recall scores; statistic computes how many times a model predicted correctly across this dataset. In this investigation, the measure variables were applied to both the training and test data sets.

The produced accuracy and recall numbers may be used to evaluate the models' overall performance when detecting (two-class) Preclinical AD and Non-preclinical AD. The suggested ML classifier approach distinguishes between individuals with positive and negative classifications based on recall and F1-score [20]. The receiver operating characteristic (ROC) graph combines the confusion matrix of each threshold based on the number of false positives accepted. The area under the curve (AUC) can also be used to help determine which technique of categorization is best. The recall is another highly helpful metric of the percentage of positive values that were accurately identified as positive out of the total number of positive values. Accuracy is the proportion of correct predictions among all analyzed cases. Figure 3 model analysis based on ROC is depicted below-

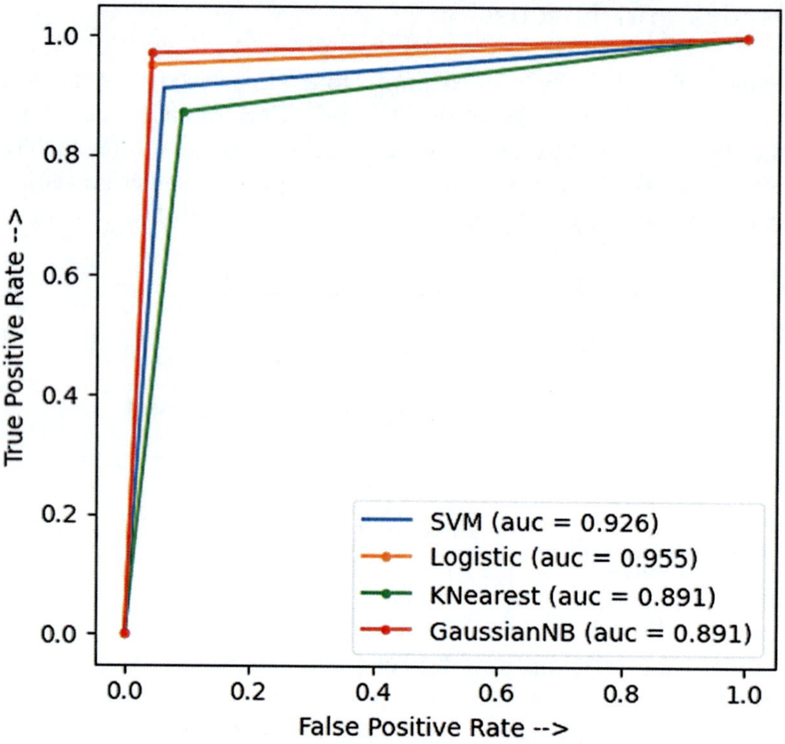

Fig. 3. ROC Representation

Figure 3 the x-axis represents the False Positive Rate (1-specificity), while the y-axis represents the True Positive Rate (Sensitivity) as calculated by plotting various thresholds (0–1). The ROC curve is used to assess the accuracy of physiologic subjects and to evaluate the diagnostic test's performance. Figures 4 and 5 provide a comparison of the suggested approach statistical measures (accuracy, precision, and recall) with machine learning models.

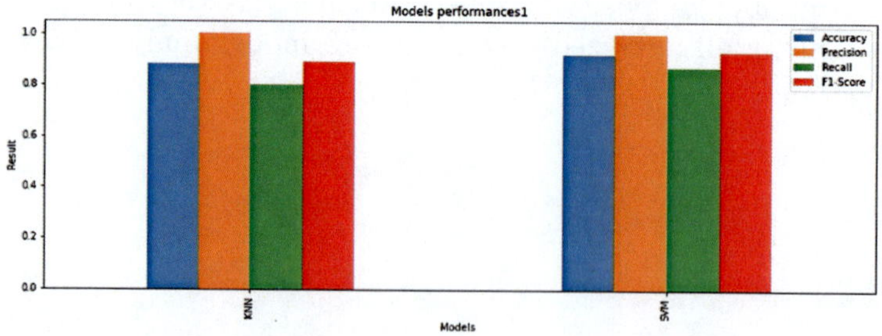

Fig. 4. Model Represent1 (KNN, SVM)

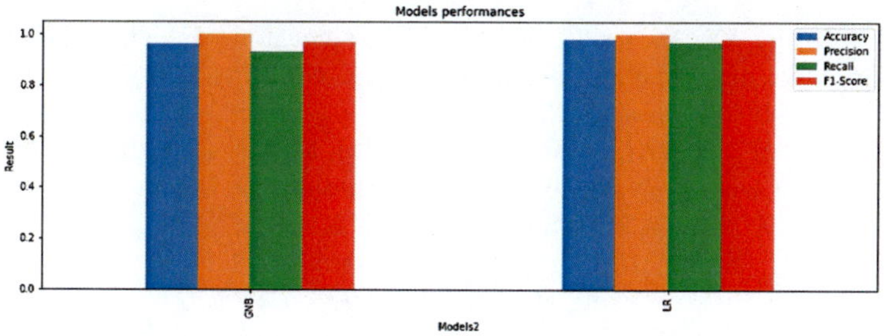

Fig. 5. Model Represent2(GNB, LR)

In, Fig. 4 and Fig. 5 the y-axis represents accuracy, precision, recall, and F1-score, whereas the x-axis represents models. It aids in determining the proper decision diagram. In this section, we compare several models in terms of precision, recall, and accuracy. Logistic Regression has a good accuracy, recall, and precision rate. Displays a graphical comparison of four model training classifiers based on their respective accuracy. The classification accuracy of KNN, SVM, GNB, and LR are 0.88, 0.92, 0.96, and 0.98, respectively. The best accuracy is predicted by LR. Generally, classification algorithms perform well in classifying patients as Preclinical or Non-Preclinical AD.

5 Conclusions

In this article, the most common classification approaches, SVM, LR, KNN, and GNB, are used to predict the progression of preclinical and non-preclinical Alzheimer's disease. After removing all DTI data, SelectkBest selected 22 features: Gender01, Marry123, ABETA1700, ABETA12, TAU80, PTAU8, AD123, AGE, EDU, PTRACCAT, APOE4. Logistic Regression has a maximum accuracy of 98%. This entire approach accurately predicts the preclinical stage of AD patients. The Classifiers listed above are independently trained and evaluated using level-based train-test split data scores. The limitation of this study is that the amount of data is insufficient. This work gives fresh information application of supervised learning paired with automated early detection. In the future, this sort of work will enhance illness prediction by merging brain MRI pictures and psychosocial traits with machine learning algorithms.

References

1. Eke, C.S., Jammeh, E., Li, X., Carroll, C., Pearson, S., Ifeachor, E.: Early detection of alzheimer's disease with blood plasma proteins using support vector machines. IEEE J. Biomed. Health Inf. **25**, 218–226 (2020). https://doi.org/10.1109/jbhi.2020.2984355
2. Billeci, L., Badolato, A., Bachi, L., Tonacci, A.: Machine learning for the classification of alzheimer's disease and its prodromal stage using brain diffusion tensor imaging data: a systematic review. Processes **8**(9), 1071 (2020). https://doi.org/10.3390/pr8091071

3. Fan, Z., Xu, F., Qi, X., Li, C., Yao, L.: Classification of Alzheimer's disease based on brain MRI and machine learning. Neural Comput. Appl. (2019). https://doi.org/10.1007/s00521-019-04495-0
4. Islam, J., Zhang, Y.: Brain MRI analysis for Alzheimer's disease diagnosis using an ensemble system of deep convolutional neural networks. Brain Inf. **5**(2), 1–14 (2018). https://doi.org/10.1186/s40708-018-0080-3
5. Almubark, I., Chang, L., Nguyen, T., Turner, R.S., Jiang, X.: Early detection of alzheimer's disease using patient neuropsychological and cognitive data and machine learning techniques. In: IEEE International Conference on Big Data (Big Data) 2019, pp. 5971–5973 (2019). https://doi.org/10.1109/BigData47090.2019.9006583
6. Lazli, L., Boukadoum, M., Mohamed, O.A.: A survey on computer-aided diagnosis of brain disorders through mri based on machine learning and data mining methodologies with an emphasis on alzheimer disease diagnosis and the contribution of the multimodal fusion. Appl. Sci. **10**(5), 1894 (2020). https://doi.org/10.3390/app10051894
7. Ahmad, F., Zulifqar, H., Malik, T.: Classification of Alzheimer disease among susceptible brain regions. Int. J. Imaging Syst. Technol. (2019). https://doi.org/10.1002/ima.22308
8. Neelaveni, J., Devasana, M.S.G.: Alzheimer disease prediction using machine learning algorithms. In: 2020 6th International Conference on Advanced Computing and Communication Systems (ICACCS), pp. 101–104 (2020). https://doi.org/10.1109/ICACCS48705.2020.9074248.
9. Yamashita, A.Y., Falcão, A.X., Leite, N.J.: The residual center of mass: an image descriptor for the diagnosis of alzheimer disease. Neuroinformatics (2018). https://doi.org/10.1007/s12021-018-9390-0
10. Castillo-Barnes, D., Su, L., Ramírez, J., Salas-Gonzalez, D., Martinez-Murcia, F.J., Illan, I.A.: (DIAN), D. I. A. N.: autosomal dominantly inherited alzheimer disease: analysis of genetic subgroups by machine learning. Inf. Fusion (2020). https://doi.org/10.1016/j.inffus.2020.01.001
11. Luckett, P.H., et al.: Modeling autosomal dominant Alzheimer's disease with machine learning. Alzheimer's Dementia **17**(6), 1005–1016 (2021). https://doi.org/10.1002/alz.12259
12. Bron, E.E., et al.: Cross-cohort generalizability of deep and conventional machine learning for MRI-based diagnosis and prediction of Alzheimer's disease. NeuroImage: Clinical **2021**, 102712 (2021). ISSN 2213-1582. https://doi.org/10.1016/j.nicl.2021.102712
13. Alshammari, M., Mezher, M.: A modified convolutional neural networks for MRIbased images for detection and stage classification of alzheimer disease. In: National Computing Colleges Conference (NCCC) 2021, pp. 1–7 (2021). https://doi.org/10.1109/NCCC49330.2021.9428810
14. Collazos-Huertas, D., Cardenas-Pena, D., Castellanos-Dominguez, G.: Instance-based representation using multiple kernel learning for predicting conversion to Alzheimer disease. Int. J. Neural Syst. (2018). https://doi.org/10.1142/s0129065718500429
15. Marzban, E.N., Eldeib, A.M., Yassine, I.A., Kadah, Y.M., for the Alzheimer's Disease Neurodegenerative Initiative: Alzheimer's disease diagnosis from diffusion tensor images using convolutional neural networks. PLoS ONE **15**(3), e0230409 (2020). https://doi.org/10.1371/journal.pone.0230409

16. Nawaz, H., Maqsood, M., Afzal, S., Aadil, F., Mehmood, I., Rho, S.: A deep feature-based real-time system for Alzheimer disease stage detection. Multimedia Tools Appl. (2020). https://doi.org/10.1007/s11042-020-09087-y
17. Xiao, R., et al.: Early diagnosis model of Alzheimer's disease based on sparse logistic regression with the generalized elastic net. Biomed. Signal Process. Control **66**, 102362 (2021). https://doi.org/10.1016/j.bspc.2020.102362
18. Kruthika, K.R., Maheshappa, H.D.: Multistage classifier-based approach for Alzheimer's disease prediction and retrieval. Inf. Med. Unlocked **14**, 34–42 (2019). https://doi.org/10.1016/j.imu.2018.12.003
19. Battineni, G., Chintalapudi, N., Amenta, F.: Machine learning in medicine: performance calculation of dementia prediction by support vector machines (SVM). Inf. Med. Unlocked **16**, 100200 (2019). https://doi.org/10.1016/j.imu.2019.100200
20. Alickovic, E., Subasi, A.: Automatic detection of alzheimer disease based on histogram and random forest. CMBEBIH **2019**, 91–96 (2019). https://doi.org/10.1007/978-3-030-17971-7-14
21. Nanni, L., Interlenghi, M., Brahnam, S., Salvatore, C., Papa, S., Nemni, R.: Comparison of transfer learning and conventional machine learning applied to structural brain MRI for the early diagnosis and prognosis of Alzheimer's disease. Front. Neurol. **11**, 576194 (2020). https://doi.org/10.3389/fneur.2020.576194
22. Tabarestani, S., et al.: Longitudinal prediction modeling of alzheimer disease using recurrent neural networks. In: 2019 IEEE EMBS International Conference on Biomedical & Health Informatics (BHI) (2019). https://doi.org/10.1109/bhi.2019.8834556
23. Shah, A., Lalakiya, D., Desai, S., Patel, V.: Early detection of alzheimer's disease using various machine learning techniques: a comparative study. In: 2020 4th International Conference on Trends in Electronics and Informatics (ICOEI), vol. 48184 (2020). https://doi.org/10.1109/icoei48184.2020.9142975
24. Ebrahimi-Ghahnavieh, A., Luo, S., Chiong, R.: Transfer learning for alzheimer's disease detection on MRI images. In: 2019 IEEE International Conference on Industry 4.0, Artificial Intelligence, and Communications Technology (IAICT), pp. 133–138 (2019). https://doi.org/10.1109/ICIAICT.2019.8784845
25. Marghalani, B.F., Arif, M.: Automatic classification of brain tumor and alzheimer's disease in MRI. Procedia Comput. Sci. **163**, 78–84 (2019). https://doi.org/10.1016/j.procs.2019.12.089
26. Moscoso, A., et al.: Prediction of Alzheimer's disease dementia with MRI beyond the short-term: implications for the design of predictive models. NeuroImage Clin. **23**, 101837 (2019). https://doi.org/10.1016/j.nicl.2019.101837
27. Khagi, B., Lee, C. G., Kwon, G.-R.: Alzheimer's disease classification from brain MRI based on transfer learning from CNN. In: 2018 11th Biomedical Engineering International Conference (BMEiCON) (2018). https://doi.org/10.1109/bmeicon.2018.8609974
28. Zhao, N., Liu, C.-C., Qiao, W., Bu, G.: Apolipoprotein E, receptors, and modulation of Alzheimer's disease. Biol. Psychiatry **83**, 347–357 (2017). https://doi.org/10.1016/j.biopsych.2017.03.003
29. Ben Ammar, R., Ben Ayed, Y.: Speech processing for early alzheimer disease diagnosis: machine learning based approach. In: 2018 IEEE/ACS 15th International Conference on Computer Systems and Applications (AICCSA), pp. 1–8 (2018). https://doi.org/10.1109/AICCSA.2018.8612831
30. Chitradevi, D., Prabha, S.: Analysis of brain sub regions using optimization techniques and deep learning method in Alzheimer disease. Appl. Soft Comput. **86**, 105857 (2020). https://doi.org/10.1016/j.asoc.2019.105857

31. Liu, L., Zhao, S., Chen, H., Wang, A.: A new machine learning method for identifying Alzheimer's disease. Simul. Model. Pract. Theory **99**, 102023 (2019). https://doi.org/10.1016/j.simpat.2019.102023
32. Rallabandi, V.P.S., Tulpule, K., Gattu, M.: Automatic classification of cognitively normal, mild cognitive impairment and Alzheimer's disease using structural MRI analysis. Inf. Med. Unlocked **18**, 100305 (2020). https://doi.org/10.1016/j.imu.2020.100305
33. Ghoraani, B., Boettcher, L.N., Hssayeni, M.D., Rosenfeld, A., Tolea, M.I., Galvin, J.E.: Detection of mild cognitive impairment and Alzheimer's disease using dual-task gait assessments and machine learning. Biomed. Signal Process. Control **64**, 102249 (2021). https://doi.org/10.1016/j.bspc.2020.102249
34. Zulfiker, M.S., Kabir, N., Biswas, A.A., Nazneen, T., Uddin, M.S.: An in-depth analysis of machine learning approaches to predict depression. Curr. Res. Behav. Sci. **2**, 100044 (2021). https://doi.org/10.1016/j.crbeha.2021.100044
35. Rohini, M., Surendran, D.: Classification of neurodegenerative disease stages using ensemble machine learning classifiers. Procedia Comput. Sci. **165**, 66–73 (2019). https://doi.org/10.1016/j.procs.2020.01.071
36. Akhund, T.M.N.U., Mahi, M.J.N., Tanvir, A.N.M.H., Mahmud, M., Kaiser, M.S.: ADEPTNESS: alzheimer's disease patient management system using pervasive sensors-early prototype and preliminary results. In: Brain Informatics: International Conference, BI 2018, Arlington, TX, USA, 7–9 December 2018, Proceedings, vol. 11, pp. 413-422 (2018). https://doi.org/10.1007/978-3-030-05587-5_39

Exploring the Link Between Brain Waves and Sleep Patterns with Deep Learning Manifold Alignment

Yosef Bernardus Wirian[1], Yang Jiang[2], Sylvia Cerel-Suhl[3], Jeremiah Suhl[3], and Qiang Cheng[1](✉)

[1] Computer Science Department, University of Kentucky, Lexington, KY 40536, USA
{yosef.wirian,qiang.cheng}@uky.edu

[2] Behavioral Science Department, University of Kentucky, Lexington, KY 40536, USA
yjiang@uky.edu

[3] Sleep Center, Lexington Veterans Affairs Medical Center, Lexington, KY 40511, USA
{cerelsuhl,jeremiah.suhl}@va.gov

Abstract. Medical data are often multi-modal, which are collected from different sources with different formats, such as text, images, and audio. They have some intrinsic connections in meaning and semantics while manifesting disparate appearances. Polysomnography (PSG) datasets are multi-modal data that include hypnogram, electrocardiogram (ECG), and electroencephalogram (EEG). It is hard to measure the associations between different modalities. Previous studies have used PSG datasets to study the relationship between sleep disorders and quality and sleep architecture. We leveraged a new method of deep learning manifold alignment to explore the relationship between sleep architecture and EEG features. Our analysis results agreed with the results of previous studies that used PSG datasets to diagnose different sleep disorders and monitor sleep quality in different populations. The method could effectively find the associations between sleep architecture and EEG datasets, which are important for understanding the changes in sleep stages and brain activity. On the other hand, the Spearman correlation method, which is a common statistical technique, could not find the correlations between these datasets.

Keywords: Deep Learning · Manifold Alignment · EEG · Sleep Architecture

1 Introduction

Nowadays, complex data usually come from different information sources and have disparate formats and representations, such as images, text, and audio. Such data are known to be multi-modal because they belong to different modalities or categories of information. However, the data of disparate sources in these datasets are not independent or unrelated. They often have certain intrinsic connection that links them together. For instance, a dataset that has images and textual captions for each image is multi-modal. The images and the captions are different types of data, but they both convey the same or similar meaning or information about the scene depicted in the image.

M. Younas et al. (Eds.): Deep-BDB 2023, LNNS 768, pp. 81–90, 2023.
https://doi.org/10.1007/978-3-031-42317-8_7

In sleep study, polysomnography (PSG) data typically contains various multi-modal datasets. They include but are not limited to electroencephalogram (EEG), electrocardiogram (ECG), and sleep architecture based on the hypnogram. The task of measuring the correlations of features within one modality is well studied and understood; however, the problem of measuring the associations between two or more modalities is unclear and often challenging. Researchers have conducted a comprehensive analysis of the relationship between PSG datasets, which are used to diagnose sleep disorders and monitor sleep quality. Picard-Deland et al. [1, 2] suggested that sleep latency and sleep spindle play a role in influencing the occurrence and intensity of nightmares in individuals. Purcell et al. [3] suggested the sleep spindle as a distinctive feature of sleep stage 2. In this study, we leveraged a novel method for exploring the relationships between PSG datasets, particularly EEG and sleep architecture datasets. This method was based on deep learning manifold alignment by Nguyen et al. [4], which performed the alignment with the supervision of the class labels. We adapted the feature alignment to an unsupervised setting where no class labels of the data were needed.

The rest of the paper is presented as follows. In Sect. 2, we provide a review of the relevant studies. In Sect. 3, we describe the datasets and preprocessing procedures, as well as the machine learning algorithm to be used. In Sect. 4, we report and analyze the results of our experiment. Finally, Sect. 5 concludes the paper by highlighting our contributions and outlining several avenues for future research.

2 Related Work

Nguyen et al. [4] introduced a method for deep manifold regularized alignment (deepManReg), which could exploit multi-modal data, such as single-cell multi-omics data, to predict the characteristics of complex biological systems. It implemented deep neural networks (DNN) for learning and aligning cross-modal manifolds, and it used the nonlinear patterns to enhance the prediction models and discover the relevant features and interactions for the characteristics. The method was demonstrated to outperform several existing methods, such as linear manifold alignment (LMA), canonical correlation analysis (CCA), and MATCHER, in predicting phenotypes via prioritizing the multi-modal features and cross-modal interactions according to their importance for the phenotypes. The effectiveness of deepManReg was demonstrated on two datasets in [4]: one with images of handwritten digits with varying attributes and the other with single-cell multi-modal data of mouse brain cells. Though the deepManReg was mainly designed for phenotype prediction in [4], we adapted it for exploring the potentially nonlinear, complex relationships between cross-modal features such as those of EEG and sleep architectures with no need of class labels in this paper.

The Spearman correlation [5] is commonly used to assess how well two variables are related in strength and direction. Unlike the Pearson correlation, which assumes a linear relationship between the variables and a normal distribution of the data, the Spearman correlation does not require these assumptions. Instead of using the actual values of the variables, the Spearman correlation calculates the degree of their association based on their ranks, thus being insensitive to the data distributions and outliers. While powerful, it does not take into account the geometrical structures of the data, thus often ignoring important inter-relationships between cross-modal variables.

There have been accumulating works linking sleep features to brain functional, structural, and pathological characteristics. The strong crosstalk between sleep and cognition has been reported [6]. New evidence has also shown strong associations between sleep oscillations (measured by EEG) and sleep-dependent memory processing. Improving sleeping health has emerged as an intervention strategy for boosting metabolic, immune, and cardiovascular systems as well as for sleep-dependent memory consolidation, and reinforcing executive processes including working memory, mood, and other cognitive functions. Recently, improving sleeping health has emerged as an intervention strategy for various diseases, including metabolic diseases, all-cause mortality, and Alzheimer's diseases [7–10]. Sleep EEG data (less noisy than daytime EEG recording) has long been used for investigating memory functions [11, 12]. For example, normal aging effects on cognition were reliably estimated from the sleep EEG data [13]; sleep EEG-based brain age index has been studied to find the association between sleep and dementia [14]. Despite intensive research and progress, the relationship between various sleep EEG features and sleep architectural characteristics, particularly spindles, remains largely unexplored. This paper seeks to tackle this unmet need by using the cross-modal feature alignment.

3 Methodology

3.1 Dataset

For this study, we obtained the datasets from the Sleep Heart Health Study (SHHS) [15, 16], which is a large-scale, multi-center, prospective cohort study of the cardiovascular consequences of sleep-disordered breathing. The datasets were available from the National Sleep Research Resource (NSRR) [17], an online repository of sleep data and tools. To simplify the analysis, we only included the data from the first clinic visit and polysomnogram of each subject, which were conducted between November 1995 and January 1998. We removed subjects who had non-numeric values in the data, such as missing or invalid entries. This resulted in 4,482 subjects out of the original 5,782 subjects. These subjects had a mean age of 62.68 years old at the time of the clinic visit, with a minimum age of 39 years old and a maximum age of 90 years old. The sex distribution of the subjects was slightly skewed towards women, with a female-to-male ratio of 1.15:1.

From the polysomnography data, we extracted two modalities: EEG and sleep architecture. We used these datasets to learn cross-modal feature alignment and explore their relationships. Both datasets were derived from the original European Data Format (EDF) signal files by using the Compumedics Profusion software. The EEG dataset contains 58 EEG biomarkers that measure the brain activity of each subject during sleep. The EEG biomarkers include the following: the average Odds Ratio Product (ORP) in all 30-s epochs of a sleep stage, which indicates the level of arousal and sleep quality; the spindle traits (density, frequency, and percentage of fast spindles) from C3 and C4 EEG channels, which reflect the cognitive functions and memory consolidation; and the average power of different spectral bands (alpha (7.33–12.0 Hz), beta (14.3–20.0 and 20.3–35 Hz), delta (0.33–2.33 Hz), gamma omega (35.3–60.0 Hz), sigma (12.33–14.0 Hz), and theta (2.67–6.33 Hz)) in all 3-s epochs for C3 and C4 channels, which represent the different stages and cycles of sleep. The biomarkers of the EEG dataset are presented in Supplementary[1].

The sleep architecture dataset is a rich source of data that captures the sleep cycle patterns of individuals who participated in type II polysomnography, a sleep study that monitors brain waves, oxygen levels, heart rate, and breathing. The dataset contains 55 extracted attributes that measure the quality and quantity of sleep, such as the number of times the individual shifted from one sleep stage to another (stage 1, 2, 3/4, and Rapid Eye Movement (REM)), the total amount of time spent in each sleep stage (and how it compares to the total sleep time), and the percentage of total sleep time that was compromised by different conditions that affect breathing and oxygenation, such as having oxygen saturation below a certain threshold, experiencing desaturation events (drops in oxygen levels), having apneas (pauses in breathing), or having hypopneas (shallow breathing). The dataset represents a comprehensive profile of the sleep architecture of the individuals, and the attributes are presented in Supplementary Table 2[2].

3.2 Deep Learning Manifold Alignment Method

Manifold alignment is a machine learning technique that can learn from multi-modal datasets [18]. It can discover the relationship between the features of different modalities by projecting them onto a common latent manifold, thereby forming a low-dimensional representation to capture the shared structure of the data. Two features are considered to share the same characteristic when their distance on the manifold is small. The correlations of features within one modality are well studied, but the associations across modalities are unclear, especially when they are nonlinear. Manifold alignment can help tackle this problem by finding a low-dimensional embedding that preserves the high-dimensional data structure and the inter-data correspondences. Although the technique works well for nonlinear problems, it can also be used in linear scenarios.

Two modal datasets with a set of p samples described in [4] as:

$$X = \{x_i\}_{i=1}^{n} \text{ and } Y = \left\{y_j\right\}_{j=1}^{m}, \tag{1}$$

where X and Y are the features of modal 1 and modal 2, respectively, n and m are the number of features of X and Y, respectively, and x_i and y_j are the ith feature of modal 1 and the jth feature of modal 2 with both of which being p-dimensional. We want to learn two mappings $f(x_i)$ and $g(y_j)$ onto the latent space with dimension $d << p$. F is the joint representation of X and Y features on the shared latent space,

$$F = \begin{bmatrix} f(X) \\ \hline g(Y) \end{bmatrix}, \tag{2}$$

where $f(X)$ is of size n x d and $g(Y)$ is of size m x d, and F is of size $(n + m)$ x d. To reveal nonlinear relationships across modalities, we perform the manifold alignment by

1 https://github.com/ybwirian/sleepmanifoldalignment/blob/main/EEGDataset.csv.

2 https://github.com/ybwirian/sleepmanifoldalignment/blob/main/SleepArchitectureDataset.csv.

adapting the deepManReg algorithm [4]. The objective function of this alignment is

$$\min_{f,g} Tr\left(F^T LF\right), s.t. F^T DF = I, \tag{3}$$

where L is the Laplacian matrix of the similarity matrix W that is of size $(n + m)$ x $(n + m)$ and defines the correspondence between every feature of X and that of Y, and D is the diagonal matrix of the row-sum of W.

We trained two DNNs to represent f and g that respectively map the EEG and sleep architecture modalities to a common latent manifold. The latent manifold was chosen to be a 10-dimensional Stiefel manifold, which constitutes a set of orthogonal matrices. Both DNNs had the same network architecture: an input layer with 4,482 nodes that corresponded to the number of subjects, two hidden layers with 512 and 64 nodes, and an output layer with 10 nodes. We trained the DNNs for 55 epochs and implemented the Adam algorithm as the optimizer. We used the k-Nearest Neighbors (kNN) algorithm with $k = 5$ to measure the similarity between the modalities in the latent manifold. A similarity square matrix with a size equal to the total number of features (i.e., 113) was obtained, where each element represented the number of common neighbors between two points. We set a threshold of 0.92 to select similar features across the modalities. The optimization of Eq. (3) followed that of [4] over the Stiefel manifold.

4 Experimental Results

The latent space generated by the manifold alignment constructed two graphs that capture the similarity between features across EEG and sleep architecture datasets. Figure 1 demonstrates how one of the features from the sleep architecture dataset, namely sleep latency, has strong nonlinear relationships with many features from the EEG dataset. Such features include the average ORP, which measures the overall regularity of sleep patterns, the Sleep Spindle traits, which reflect the bursts of brain activity during non-REM sleep, and the average spectral band power, which indicates the amount of energy in different frequency ranges. These results are consistent with those in the previous study on sleep and nightmares by Picard-Deland et al. [1, 2] that suggested sleep latency and sleep spindle play a role in influencing the occurrence and intensity of nightmares in individuals.

The graph in Fig. 2 demonstrates another set of high associations between the features of EEG and sleep architecture datasets in the aligned latent space. We can observe that two features about the sleep architecture, which measure the percentage of total sleep duration with oxygen saturation below 75%, have a strong nonlinear relationship with the average alpha spectral band power from the EEG dataset. Another two features from the sleep architecture dataset, which count the number of shifts from/to stage 2 per hour of sleep, strongly correlate with the sleep spindle feature from the EEG dataset. This finding is in line with the existing literature by Purcell et al. [3] that suggested sleep spindle as a distinctive feature of sleep stage 2. These relationships between cross-modality features were already known, thus substantiating the usefulness of our approach. There are other relations in Figs. 1 and 2, which appear to be novel. We will explore these newly found relationships, particularly under various disease conditions, in our future studies.

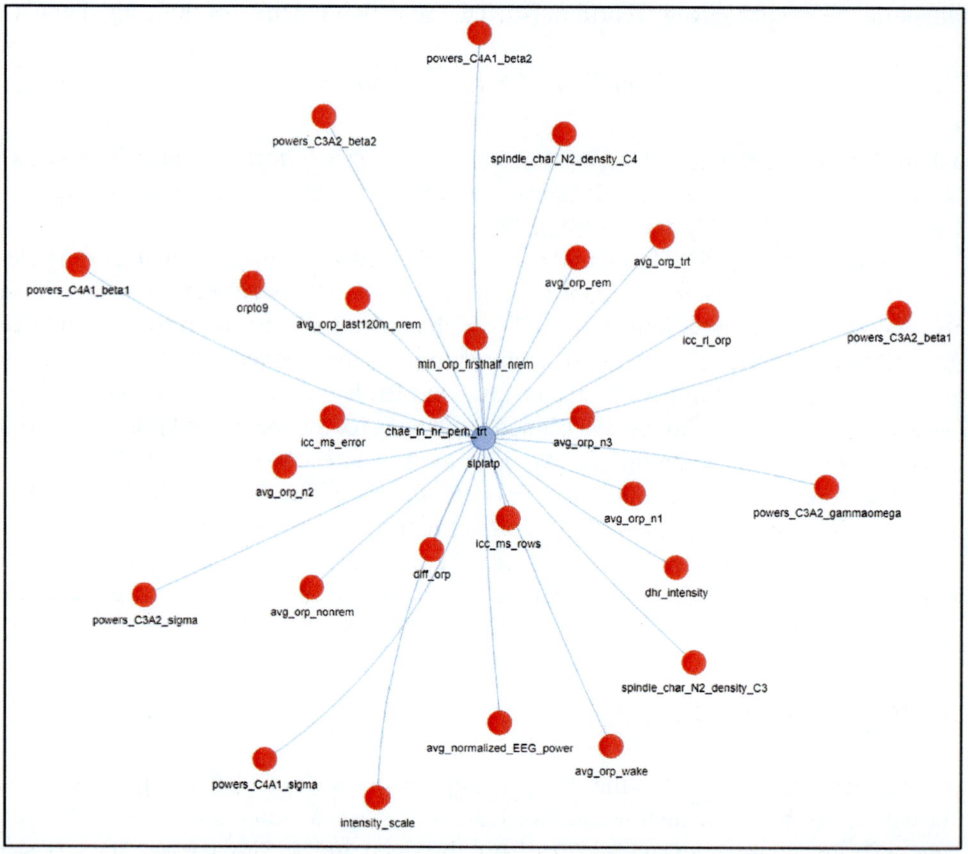

Fig. 1. The first network shows the relationships across EEG and Sleep Architecture datasets. The red circle represents the EEG dataset, and the blue circle represents the Sleep Architecture dataset. The distance between a pair of nodes is inverse proportional to the strength of their association; the shortest the distance the stronger the relation.

slplatp; Sleep Latency: the interval between lights-out/in-bed time and sleep onset time from type II polysomnography
powers_C3A2_sigma; Powers C3/A2 (mV^2) - Sigma (12.33-14.0)
powers_C3A2_beta1; Powers C3/A2 (mV^2) - Beta1 (14.3-20.0)
powers_C3A2_beta2; Powers C3/A2 (mV^2) - Beta2 (20.3-35)
powers_C3A2_gammaomega; Powers C3/A2 (mV^2) - Gamma Omega (35.3-60.0)
powers_C4A1_sigma; Powers C4/A1 (æ¦V^2) - Sigma (12.33-14.0)
powers_C4A1_beta1; Powers C4/A1 (æ¦V^2) - Beta1 (14.3-20.0)
powers_C4A1_beta2; Powers C4/A1 (æ¦V^2) - Beta2 (20.3-35.0)
avg_orp_wake; Average ORP During Wake

avg_orp_nonrem; Average ORP During non-REM
avg_orp_n1; Average ORP During N1
avg_orp_n2; Average ORP During N2
avg_orp_n3; Average ORP During N3
avg_orp_rem; Average ORP During REM
avg_org_trt; Average ORP Over Total Recording Time
avg_orp_last120m_nrem; Avg ORP Last 120 min NREM
min_orp_firsthalf_nrem; Min ORP First Half NREM
diff_orp; Change in ORP across the night
intensity_scale; Arousal Characteristics - Intensity Scale
dhr_intensity; Arousal Characteristics - dHR/ Intensity
chae_in_hr_perh_trt; Arousal Characteristics - Change in HR per hr of TRT
avg_normalized_EEG_power; Average Normalized EEG Power
icc_rl_orp; Icc R/L Orp
icc_ms_rows; Breakdown of ICC components - MS Rows
icc_ms_error; Breakdown of ICC components - MS Error
orpto9; Orp-9
spindle_char_N2_density_C3; Spindle Characteristics (Stage N2) - Density (C3)
spindle_char_N2_density_C4; Spindle Characteristics (Stage N2) - Density (C4)

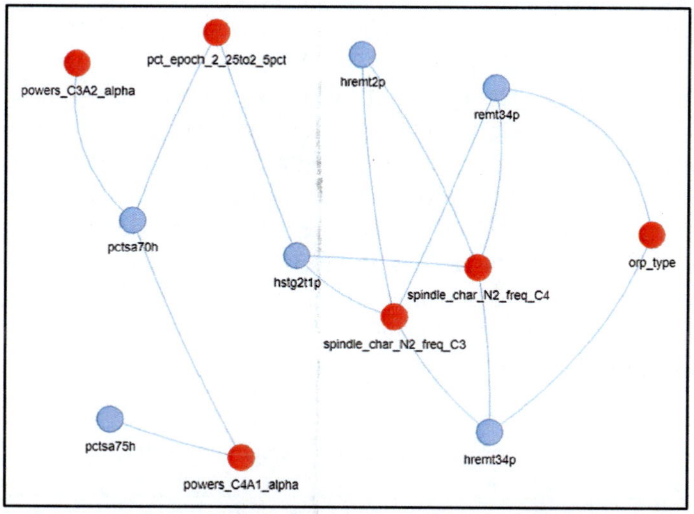

Fig. 2. The second network shows the relationships across EEG and Sleep Architecture datasets. The red circle represents the EEG dataset, and the blue circle represents the Sleep Architecture dataset. The distance between a pair of nodes is inverse proportional to the strength of their association; the shortest the distance the stronger the relation.

hremt34p; Number of shift (sleep stages shift) from REM sleep to stage 3/4 sleep per hour of sleep from type II polysomnography
remt34p; Total number of shift (sleep stages shift) from REM sleep to stage 3/4 sleep from type II polysomnography
hstg2t1p; Number of shift (sleep stages shift) from stage 2 to stage 1 sleep per hour of sleep from type II polysomnography
hremt2p; Number of shift (sleep stages shift) from REM sleep to stage 2 sleep per hour of sleep from type II polysomnography
pctsa70h; Percent total sleep duration with below 70% oxygen saturation from type II polysomnography
pctsa75h; Percent total sleep duration with below 75% oxygen saturation from type II polysomnography
powers_C3A2_alpha; Powers C3/A2 (mV^2) - Alpha (7.33-12.0)
powers_C4A1_alpha; Powers C4/A1 (æ¦V^2) - Alpha (7.33-12.0)
spindle_char_N2_freq_C3; Spindle Characteristics (Stage N2) - Frequency (C3)
spindle_char_N2_freq_C4; Spindle Characteristics (Stage N2) - Frequency (C4)
orp_type; ORP Type
pct_epoch_2_25to2_5pct; Distribution of epochs in 2.25-2.5% ORP decile

To compare the performance of manifold alignment with common method, Spearman correlation was calculated to measure the relationship between two features from the sleep architecture dataset and sleep spindle features from the EEG dataset. The two sleep architecture features were sleep latency and the count of the number of shifts from/to stage 2 per hour of sleep. The results of Spearman correlation in Table 1 showed that only one feature from the sleep architecture dataset, namely, the count of the number of shifts from/to stage 2 per hour of sleep, had a statistically significant correlation (p-value < 0.05) with any of the sleep spindles features from the EEG dataset. However, none of the features from the sleep architecture dataset had a noticeable correlation with any of the sleep spindle features from the EEG dataset because the absolute value of the Spearman correlation coefficients were all very small. This suggests that Spearman correlation was unable to capture the underlying relationship between these features; in contrast, our approach could find meaning nonlinear relations.

Table 1. Spearman Correlation Result.

Sleep Architecture	EEG	Spearman Correlation	P-value
slplatp	spindle_char_N2_density_C3	−8.078E-03	5.887E-01
slplatp	spindle_char_N2_density_C4	−3.796E-03	7.995E-01
hremt34p	spindle_char_N2_freq_C3	6.861E-03	6.461E-01
hremt34p	spindle_char_N2_freq_C4	4.607E-03	7.578E-01
hremt2p	spindle_char_N2_freq_C3	8.707E-02	5.257E-09
hremt2p	spindle_char_N2_freq_C4	7.989E-02	8.545E-08
hstg2t1p	spindle_char_N2_freq_C3	−2.484E-02	9.636E-02
hstg2t1p	spindle_char_N2_freq_C4	−1.718E-02	2.502E-01

5 Conclusion and Future Work

In this study, we applied a novel deep learning manifold alignment method to explore the complex relationships between different polysomnography (PSG) datasets, with a focus on electroencephalography (EEG) and sleep architecture datasets. The results of our analysis were consistent with the findings of previous studies that utilized PSG datasets to diagnose various sleep disorders and monitor sleep quality in different populations. The proposed method was able to effectively find the associations between EEG and sleep architecture datasets, which are essential for understanding the dynamics of sleep stages and brain activity. In contrast, the Spearman correlation method, which is a commonly used statistical technique for measuring nonlinear relationships between a pair of variables, failed to capture the correlations between these datasets.

The current results will inform future longitudinal studies for mental health improvement, both while awake and during nap or night sleep in during brain aging. One of the potential avenues for future research is to explore the relationship between the proposed method and other types of PSG datasets that measure different physiological signals, such as electrocardiogram (ECG) and electromyography (EMG). These datasets could provide additional insights into the sleep quality and patterns of the subjects. Furthermore, this method could be utilized as a model for the early detection of various brain diseases that affect the cognitive and motor functions of the patients, such as Alzheimer's disease, Parkinson's disease, and stroke. This could help in providing timely diagnosis and intervention for these conditions.

Acknowledgements. The Sleep Heart Health Study (SHHS) was supported by National Heart, Lung, and Blood Institute cooperative agreements U01HL53916 (University of California, Davis), U01HL53931 (New York University), U01HL53934 (University of Minnesota), U01HL53937 and U01HL64360 (Johns Hopkins University), U01HL53938 (University of Arizona), U01HL53940 (University of Washington), U01HL53941 (Boston University), and U01HL63463 (Case Western Reserve University). The National Sleep Research Resource was supported by the National Heart, Lung, and Blood Institute (R24 HL114473, 75N92019R002).

The study is partially supported by NIH R21 AG070909-01, P30 AG072946-01, and R01 HD101508-01.

References

1. Picard-Deland, C., Carr, M., Paquette, T., Nielsen, T.: Sleep spindles are altered in early- but not late-onset nightmare recallers. Sleep Med. **52**, 34–42 (2018). https://doi.org/10.1016/j.sleep.2018.07.015
2. Picard-Deland, C., Carr, M., Paquette, T., Saint-Onge, K., Nielsen, T.: Sleep spindle and psychopathology characteristics of frequent nightmare recallers. Sleep Med. **50**, 113–131 (2018). https://doi.org/10.1016/j.sleep.2017.10.003
3. Purcell, S., Manoach, D., Demanuele, C., et al.: Characterizing sleep spindles in 11,630 individuals from the national sleep research resource. Nat. Commun. **8**, 15930 (2017). https://doi.org/10.1038/ncomms15930
4. Nguyen, N.D., Huang, J., Wang, D.: A deep manifold-regularized learning model for improving phenotype prediction from multi-modal data. Nat. Comput. Sci. **2**(1), 38–46 (2022)

5. Spearman, C.: The proof and measurement of association between two things. Am. J. Psychol. **15**(1), 72–101 (1904). https://doi.org/10.2307/1412159
6. Simon, K.C., Nadel, L., Payne, J.D.: The functions of sleep: a cognitive neuroscience perspective. Proc. Natl. Acad. Sci. **119**(44), e2201795119 (2022)
7. Leary, E.B., et al.: Association of rapid eye movement sleep with mortality in middle-aged and older adults. JAMA Neurol. **77**, 1241–1251 (2020)
8. Winer, J.R., et al.: Sleep disturbance forecasts β-amyloid accumulation across subsequent years. Curr. Biol. **30**, 4291–4298 (2020)
9. Xiong, X., Hu, T., Yin, Z., Zhang, Y., Chen, F., Lei, P.: Research advances in the study of sleep disorders, circadian rhythm disturbances and Alzheimer's disease. Front. Aging Neurosci. **14**, 930 (2022)
10. Zhang, Y., et al.: Sleep in Alzheimer's disease: a systematic review and meta-analysis of poly-somnographic findings. Transl. Psychiatry **12**(1), 1–12 (2022)
11. Djonlagic, I., et al.: Associations between quantitative sleep EEG and subsequent cognitive decline in older women. J. Sleep Res. **28**(3), e12666 (2019)
12. Geng, D., Wang, C., Fu, Z., Zhang, Y., Yang, K., An, H.: Sleep EEG-based approach to de-tect mild cognitive impairment. Front. Aging Neurosci. **14**, 865558 (2022)
13. Sun, H., et al.: Brain age from the electroen-cephalogram of sleep. Neurobiol. Aging **1**(74), 112–120 (2019)
14. Ye, E., et al.: Association of sleep electroencephalography-based brain age index with dementia. JAMA Netw. Open **3**(9), e2017357 (2020)
15. Zhang, G.Q., et al.: The National Sleep Research Resource: towards a sleep data commons. J. Am. Med. Inform. Assoc. **25**(10), 1351–1358 (2018). https://doi.org/10.1093/jamia/ocy064
16. Quan, S.F., et al.: The Sleep Heart Health Study: design, rationale, and methods. Sleep **20**(12), 1077–1085 (1997). PMID: 9493915
17. NSRR Homepage: https://sleepdata.org/
18. Ham, J.H., Lee, D.D., Saul, L.K.: Learning high dimensional correspondences from low dimensional manifolds. In: Proceedings of the Twentieth International Conference on Machine Learning (ICML-2003) (2003)

Machine Learning and Commercial Systems

YOLOv5 for Automatic License Plate Recognition in Smart Cities

Abir Raza(✉), Elarbi Badidi, Basma Badidi, and Sarah Al Zahmi

Department of Computer Science and Software Engineering, College of Information Technology, UAE University, P.O. Box 15551, Al-Ain, United Arab Emirates
{202190274,ebadidi}@uaeu.ac.ae

Abstract. Recognition and extraction of license plate information from still images or videos are the basis of modern traffic and security systems. Automatic License Plate Recognition (ALPR) is transforming public safety and transportation in many ways. Such license plate recognition systems enable advanced solutions for toll roads, offer significant operational cost savings through automation, and even open up new market opportunities, such as license plate readers mounted on police vehicles. In this work, we implement license plate recognition based on convolutional neural networks to achieve high accuracy. You Only Look Once (YOLO) algorithm is employed for detection, while, an Optical Character Recognition (OCR) technique is used for text recognition. We use four variants of the YOLOv5 for license plate detection and the EasyOCR for license plate recognition. The results show that YOLOv5x (extra-large) achieves a Mean Average Precision (mAP) of nearly 82% in detecting vehicles and license plates in an image. Furthermore, the recognition of the letters on license plates is achieved with a confidence score of nearly 65%.

Keyword: Object detection · Object recognition · Convolutional neural networks · Deep learning

1 Introduction

Technologies for intelligent transportation systems, smart vehicles, innovative healthcare, and smart cities will continue transforming many aspects of human life. As a result, automatic license plate recognition technologies have become part of our daily activities. Moreover, the concept of ALPR promises to contribute to various use cases without human intervention. ALPR is a computer vision technique that enables devices to quickly and automatically read license plates on vehicles without human intervention. ALPR is used to accurately capture and identify license plates through video or still camera images.

Modern traffic and public safety systems recognize and extract license plate information from still images or videos. Such license plate recognition systems enable modern solutions for toll roads that provide tremendous operational cost savings through automation. In recent years, many police departments worldwide have installed ALPR systems

M. Younas et al. (Eds.): Deep-BDB 2023, LNNS 768, pp. 93–105, 2023.
https://doi.org/10.1007/978-3-031-42317-8_8

in police vehicles for various law enforcement purposes, including traffic control, parking management, access control to secure areas, toll booths, homeland security, and counter-terrorism [1].

OpenALPR is a widely used automatic license plate recognition library based on optical character recognition of images or videos of license plates. Typical ALPR systems include cameras, processing units, and video analysis algorithms. These systems can capture license plates even at night thanks to infrared illumination, allowing the ALPR system to operate around the clock. The ALPR camera captures an image (video stream or photo) containing a license plate. The license plate is then recognized using machine learning and computer vision techniques, primarily object recognition techniques. Finally, an OCR software is applied to the recognized area of the license plate to render the license plate in text format, which is then stored in a database for integration with other systems.

2 Applications of ALPR

2.1 Use Cases of ALPR

Automatic license plate recognition is essential for a variety of applications. For example, vehicle detection, identification, or localization is essential for innovative city applications.

Law Enforcement. Police use ALPR for law enforcement, e.g., to check whether a vehicle is registered or to identify vehicles associated with any traffic violations [1]. Real-time license plate recognition allows law enforcement to identify vehicles and track their location.

Parking Management. Efficient parking management requires intelligent solutions capable of recognizing individual vehicles [2]. Therefore, automatic license plate recognition and detection is the answer to efficient parking management [3]. With ALPR, each car is registered by a license plate, which enables the automatic management of parking spaces in parking garages. This way, parking lot users are spared the stress of ticket management, time recording, the risk of inaccurate ticket payment, and penalties for lost tickets. Such automated monitoring can also be used in the event of a dispute. Such parking systems track every vehicle in the facility and ensure complete control.

Traffic Management. ALPR can be used in cities to detect speeding, reckless driving, or other accidents. ALPR provides solutions to measure and analyze area-based traffic data of a specific area or even an entire city. On a larger scale, traffic management provides a better understanding of traffic congestion for better planning [4].

Safety in Commercial Areas. Unauthorized parking often occurs in commercial areas, leading to problems with legitimate parking or suspicious activities. These security risks can be addressed with ALPR technology by ensuring that only authorized vehicles use the parking lots.

Tollbooth Records. Manual management of toll booths on highways is still essential in many countries. Often, toll booths use different technologies for autonomous toll booth management. For example, on major expressways, ALPR enables agencies to receive license plates to pay tolls by mail or automatically rather than stopping and paying at a manually operated toll booth. ALPR thus enables efficient management of toll booths, reduces the required operating time, and thus increases productivity [5].

2.2 Object Detection with Deep Learning Techniques

Computer vision includes the tasks of image classification, object localization, and object detection in images. Image classification involves labeling an image, while object localization involves drawing bounding boxes around identified objects in an image. Object detection is more complex and combines classification and localization.

A bounding box is drawn around an object in an image to localize an object or demarcate the Region of Interest (RoI). Whereas, Intersection over Union (IoU) is a YOLO measure indicating how many two bounding boxes intersect. It is the overlap between the ground truth and the predicted bounding box [7]. The YOLO architecture predicts multiple bounding boxes per grid cell. The bounding boxes possessing the highest IoU with the ground truth are selected. The Non-Maximum Suppression (NMS) technique is used to suppress all unwanted bounding boxes with lower probabilities. We have used the pre-defined bounding box algorithm of YOLO to identify license plates in a vehicle image.

Algorithms for image classification return the categories of objects identified in an image. Object detection and recognition are essential for many critical applications, such as autonomous driving, video surveillance, and facial recognition. The algorithms strive to detect and identify the different instances of objects of a particular class (e.g., vehicles, people, or buildings) in digital images and videos. An axis-aligned bounding box indicates the scale and position of each object. The performance of an object recognition model is evaluated based on the accuracy and recall in each of the best-fitting bounding boxes for known objects in the image.

Recent powerful Deep Learning models include the Region based Convolutional Neural Networks (R-CNN) model family (R-CNN, Fast R-CNN, Faster R-CNN) and the YOLO model family (YOLO, YOLOv2, YOLO9000, YOLOv3, YOLOv4, YOLOv5) [6]. The YOLO algorithm is a fast and efficient convolutional neural network designed for real-time object detection and classification [7]. The algorithm consists of 24 convolutional layers and two fully connected layers and operates based on Rectified Linear Unit (ReLU). YOLO can detect objects in a single forward propagation through a convolutional neural network to make predictions, and this means that complete detection is done in a single execution. As the input passes through the neural network, the image is divided into grids, and each grid produces a bounding box of class probabilities, as in YOLO.

One of the latest versions of YOLO is YOLOv5, which is based on the PyTorch framework and surpasses previous versions in terms of flexibility [8].

3 Related Work

Over the last few years, several efforts investigated the different solutions concerning automatic license plate recognition. Systematic literature reviews on the subject are presented in [9, 10]. In these reviews, three main aspects are addressed: (i) the existing technology and its progress over time; (ii) the factors influencing its performance accuracy in different contexts; the applicability of the technology; and (iii) its practical implications from the user's perspective.

Earlier ALPR systems used open-source platforms such as OpenCV and Python [11, 12]. Some earlier solutions used basic image processing methods to recognize Arabic license plates [13, 14]. Luvizon et al. [15] used the commercially available SnooperText for license plates recognition. License plates are tracked across all frames using a combination of Kanade-Lucas-Tomas (KLT) and Scale-Invariant Feature Transform (SIFT). The system achieved an accuracy of 87% and a recall of 0.92 on a dataset of 75 vehicle sequences. Later, Kumari et al. [16] presented an ALPR technique based on the K-nearest neighbor algorithm. After image acquisition and preprocessing, a vector of possible license plate characters is computed. The character with the most significant occurrence is selected and matched with the license plate pattern. Moreover, Awalgaonkar et al. [17] proposed a deep learning-based approach for Indian license plate recognition. They used SSD-based MobileNet, a lightweight CNN for plate detection, and Tesseract OCR and Easy OCR or license plate recognition.

After the introduction of YOLO by Joseph Redmon in 2015 [7], Spanu et al. [22] developed a multiclass ALPR solution based on YOLO. The significance of the developed model lies in scenario analysis for traffic density and congestion detection, license plate recognition, and vehicle type identification for a sustainable smart city. The authors in [18] developed a real-time ALPR system using Fast-YOLO, YOLOv2, and CR-NET models. They used two separate Convolutional Neural Networks (CNNs) to train and detect vehicles and license plates. Kim et al. [21] introduced an AI camera for recognizing Korean license plates by integrating front-end and back-end intelligence. They used three CNN models, R-Net for detecting license plates, V-Net for license plate coordinates, and CNET for alphanumeric character recognition.

Ali Tourani et al. [24] designed an ALPR system for Iranian license plates using two sequential YOLOv3 deep networks. For better detection accuracy and precise IoU, they changed the last convolutional layer of YOLOv3 to only one class of Persian license plates. The authors in [20] presented a framework inspired from YOLOv3 in which a CNN induces the text recognition process. The feature vectors, generated from the license plate features, are then used as input to the long short-term memory (LSTM) layer. The Taiwan application-oriented license plate (AOLP) dataset evaluates the license plate recognition system. The authors have proposed a real-time application using YOLOv3 to detect license plates under unsuitable conditions, regardless of angle and brightness, with 80% accuracy on an Indian dataset of 2500 images [25]. Elnashar et al. [26] proposed a deep learning model to detect and classify Egyptian license plates using YOLOv3 by. They used one-shot learning to classify the license plates based on the Siamese network and a multi-task learning model. Lee et al. [27] experimented with the training parameters of YOLOv3 to investigate a model's design. They used a multilevel (2-level and 3-level) factorial design of experiments (DOE) to study the interactions between the training parameters and optimize the detection of Malaysian license plates. For ALPR systems, DOEs allow tuning the performance of YOLOv3.

Also, Sung et al. [28] presented a computationally lighter ALPR system. They utilized a Single Shot MultiBox Detector (SSD) for license plate detection and YOLOv4 for character recognition. They evaluated the performance of the system using the KETI-ALPR dataset. Other efforts designed and developed automatic vehicle tracking systems based on YOLOv4 for plate detection and Tesseract OCR for character recognition [29, 32]. One of the systems has an accuracy of almost 89% on the Google Open Images dataset.

An Edge- AI-based real-time ALPR system is proposed by Lin et al. for Taiwanian license plates [31]. The architecture uses YOLOv4-Tiny, a modified YOLOv4 character recognition, and an auxiliary logical judgment system for license plate recognition. YOLOv4 and MobileNet-SSD have been implemented for Vietnamese vehicles also [30]. The research demonstrates that MobileNet-SSD outperforms YOLOv4 due to its speed and accuracy.

CR-NET was used to recognize Brazilian and Pakistani license plates [19, 23]. Montazzolli et al. [19] used Fast-YOLO and evaluated their proposed method using the SegPlate database (SSIG) and the UFPR-ALPR dataset. The authors found that impressive results were obtained with both the datasets. Usama et al. [23] proposed an ALPR for automatic toll deductions using YOLOv4, Tiny YOLOv4, YOLOv3, Tiny YOLOv3, YOLOv2, and Faster RCNN. Their experiments on a customized dataset of 10,000 images showed that YOLOv4 scored the highest in precision, recall, F1 score, and mAP.

With the advancement of YOLO models over time, Batra et al. [33] and Raj et al. [34] employed YOLOv5 for recognizing Indian license plates. The authors made use of an LSTM-based OCR for Microsoft COCO dataset, Google Open Images, and Indian license plates data [33]. Whereas, a CNN is used in combination with YOLOv5 in the later research [34]. A license plate image is segmented and fed into the CNN. The character labels generated by the CNN are merged to estimate the license plate number. The proposed system achieves a mAP of 88.4%. Table 1 summarizes the literature review.

4 Experimentation and Results

4.1 Methodology

There are two approaches to ALPR: single-stage and multistage processing. Unlike a conventional ALPR system, our proposed pipeline includes two-stage detection with YOLO and character recognition with OCR.

YOLO's unified architecture is extremely fast and presents object detection as a single regression problem, directly from image pixels to bounding box coordinates and class probabilities [7]. In just one evaluation, a single neural network directly predicts bounding boxes and class probabilities from complete images. YOLO makes more localization errors than state-of-the-art detection systems but predicts fewer false positives in the background. Therefore, it outperforms other detection methods, such as DPM and R-CNN.

An OCR algorithm converts handwritten, typed, or printed text into a machine-readable format. It is built on PyTorch and comes as a Python library.

We have used the pipeline depicted in Fig. 1 for license plate detection and recognition. Our algorithm starts by detecting a vehicle in a given image, and then the license plate is detected and cropped out of the image for character recognition. The cropped license plate is processed and used as input for the character recognition phase in the next step.

Figure 2 shows the elaborated system architecture. First, a vehicle is detected in a given image using YOLOv5 to draw a bounding box around the vehicle. Then, the vehicle's license plate is located by delineating the boundaries (dimensions) of the license plate. YOLOv5 provides numerous models with varied configurations and parameter sizes. In our research, four variants of YOLOv5 are trained and tested, namely, YOLOv5m (medium), YOLOv5n (nano), YOLOv5s (small), and YOLOv5x (extra-large). The results of each algorithm are compared to get the best approximation of the detection. After that, we used EasyOCR for the recognition phase. In our case, EasyOCR extracts text information from the cut-out license plates.

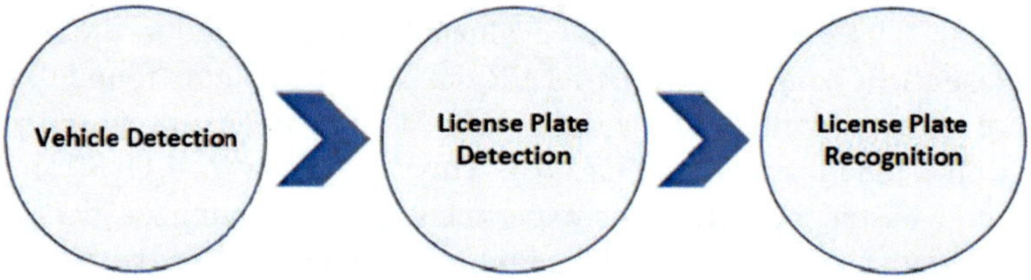

Fig. 1. Proposed pipeline for license plate detection and recognition

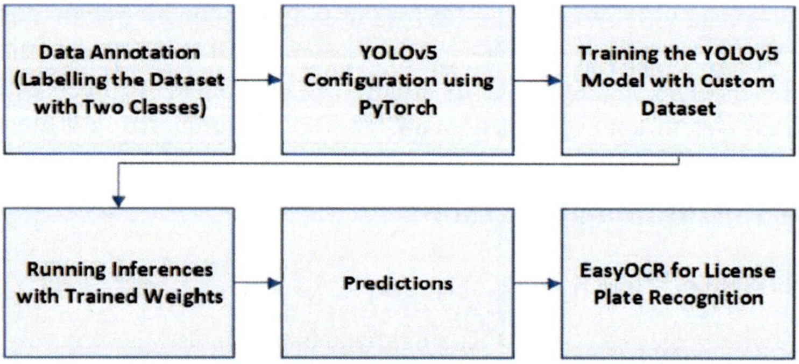

Fig. 2. System architecture

We used an open-source license plate dataset from Kaggle to train, validate, and evaluate our approach. The dataset contains 300 images of vehicles from different countries, and the license plates have different font sizes, styles, symbols, and even colors. The images in the dataset are labeled using Roboflow (https://roboflow.com/). 70% of the dataset is used for training the models, 21% is used for validation, and 9% is provided for training performance evaluation.

Table 1. Comparative analysis of the literature review

Reference	Dataset	Detection Approach	Recognition Approach	Recognition Accuracy
[18]	UFPR-ALPR, SegPlate SSIG	YOLOv2, Fast-YOLO	CR-NET	78.33% with UFPR-ALPR and 93.53% with SSIG
[21, 22]	OpenALPR, AOLP-RP, Korean LPs	YOLOv2, Mask-RCNN,	R-Net, V-Net, and C-Net [22], CRNN [23]	98.16% mAP [21], CNN = 99.37% Mask-RCNN = 97.5%. CRNN = 94.5% [22]
[20, 24, 25, 27]	AOLP, Indian, Iranian, and Malaysian LPs	YOLOv3	CRNN [21], YOLOv3 [25], OpenCV [26]	86% mAP [20], 95.05% mAP [24], 90% mAP [25], 99% mAP [27]
[26]	Egyptian LPs	YOLOv3	Siamese Network	99.2% for standard LPs
[28, 29, 32]	KETI-ALPR, Google Open Images	YOLOv4, ImageAI	YOLOv4 Tesseract OCR	98.36% mAP [28], 89%mAP [32]
[31]	Taiwan LPs	YOLOv4-Tiny	Modified YOLOv4	Day = 97% mAP Night = 95% mAP
[30]	Vietnamese LPs	MobileNet, SSD and YOLOv4	-	MobileNet-SSD = 40 FPS, YOLOv4 = 7.2 FPS
[19, 23]	DVLPD, SSIG and UFPR-ALPR	YOLOv2, YOLOv3, YOLOv4, FasterRCNN, Tiny YOLOv3, Tiny YOLOv4, Fast-YOLO	CR-NET	98.8% mAP
[33, 34]	Google Open Images and Indian LPs	YOLOv5	EasyOCR [33], LeNet [34]	87.2% mAP [33], 98.3%mAP [34]

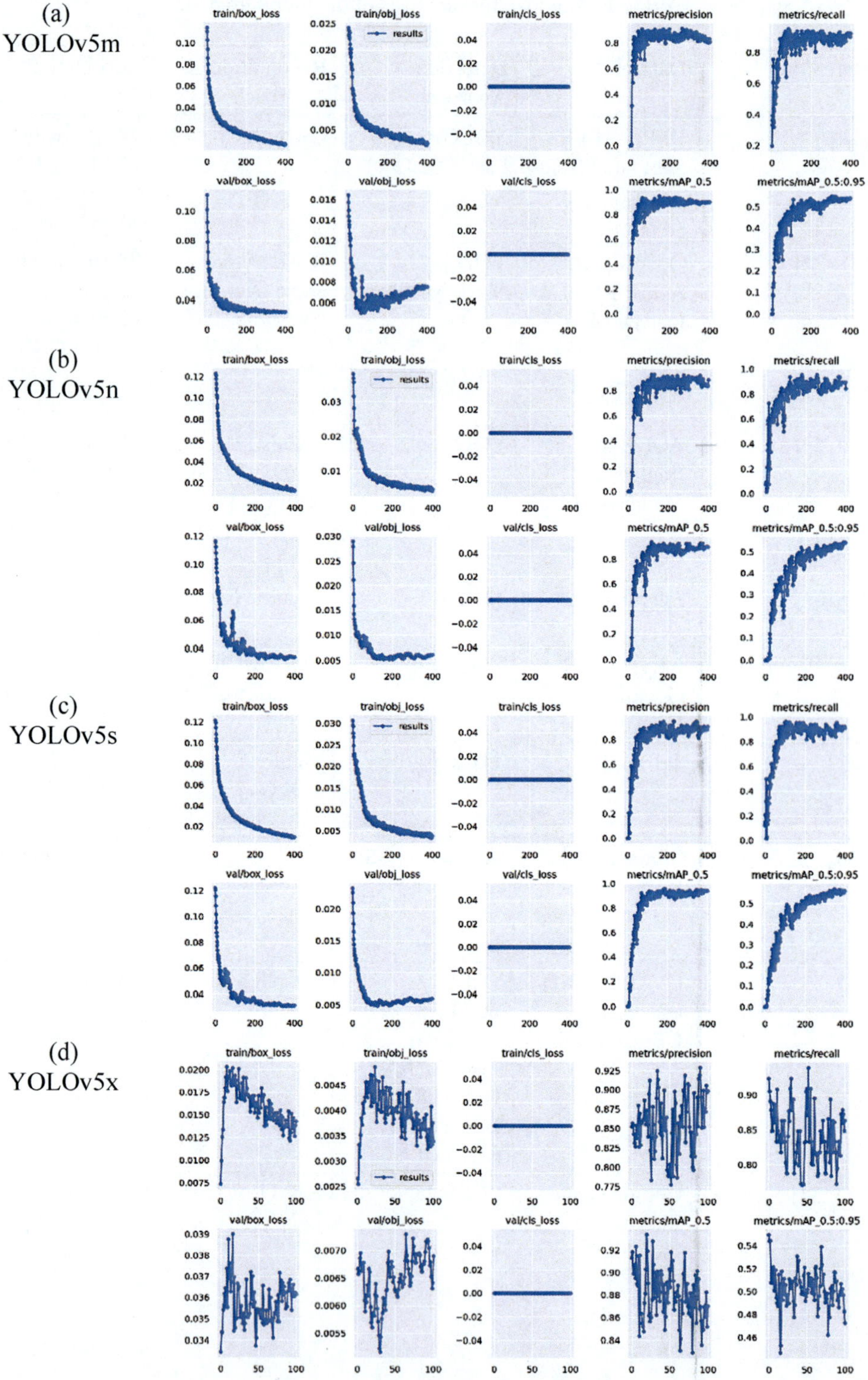

Fig. 3. Graphs of Precision, Recall, and mAP for YOLOv5

4.2 Results

After the training process, we evaluated the performance of the four models using several metrics, including Precision, Recall, and mAP@ [.5:.95], when IoU is 0.5 (50%) and 0.95 (95%). Figure 3 (a), (b), (c), and (d) show the curves of the metrics throughout the training for the different models of YOLOv5 that are trained, namely, YOLOv5m, YOLOv5n, YOLOv5s, and YOLOv5x.

The training results confirm the effectiveness of our approach in correctly recognizing license plates from the Kaggle dataset. Table 2 compares the different YOLOv5 implementations in terms of their Precision, Recall, and mAP.

Although YOLOv5x is computationally expensive, but it provides the best detection accuracy among the four variants of YOLOv5. Figure 4(a) compares the license plate detection accuracy (mAP@0.5) with four implementations of YOLOv5. Based on this analysis, we apply OCR to the license plates detected by the YOLOv5x version to optimize the results. We used EasyOCR to read only English text and numeric digits from the license plate images.

Table 2. Precision, Recall, and mAP of YOLOv5 implementations

	YOLOv5m	YOLOv5n	YOLOv5s	YOLOv5x
Precision	80%	90%	89%	93%
Recall	90%	90%	90%	99%
mAP@0.5	95%	95.5%	95.5%	99%

Figure 4(b) shows the confidence score for recognizing each license plate detected by YOLOv5x. The results show that our system achieves a 30% to 100% confidence level.

In optical character recognition, a Confidence Score indicates how confident the underlying OCR algorithm is in extracting the correct value. An OCR system provides character-level and word-level confidence scores that lie in the range of 0 to 100. In this work, we evaluate the confidence scores on the mAP at different IoU thresholds.

It is clear from the experiments that EasyOCR is quite efficient in recognizing text with different orientations. We tested EasyOCR with random license plates from the United Arab Emirates (UAE), but less than 40% is achieved in recognizing Arabic alphabets and digits. The main reason behind this is that each UAE emirate has a license plate with different colors, styles, and calligraphical fonts.

Sr. #	Vehicle Image	LP Detection Accuracy with YOLOv5m	LP Detection Accuracy with YOLOv5n	LP Detection Accuracy with YOLOv5s	LP Detection Accuracy with YOLOv5x
1		0.96	0.94	0.95	0.97
2		0.94	0.95	0.91	0.97
3		0.95	0.93	0.93	0.96
4		0.93	0.94	0.95	0.95
5		Nil (LP not detected)	Nil (LP not detected)	Nil (LP not detected)	0.80
6		0.95	0.91	0.94	0.98
7		0.91	0.89	0.94	0.94
8		0.91	0.86	0.84	0.97
9		0.95	0.94	0.92	0.95
10		0.95	0.95	0.92	0.96
11		0.92	0.92	0.92	0.96
12		0.76	0.85	0.82	0.95

(a)

Sr. #	Vehicle Image	LP Recognition with YOLOv5x		
		Cropped LP	Recognized Text	Confidence Score
1			'PGoMN112'	0.787
2			'MHO'	0.622
3			'Czi7 KoD'	0.6211
4			'DL49 AK49'	1.0
5			'MC' '09'	0.1315 0.9971
6			'BYOND'	0.4354
7			'01954521'	0.147
8			'H90Z FKL'	0.163
9			'SDN7484U'	1.0
10			'M9060.90K'	0.7589
11			'5220.89563'	0.3289
12			'172' 'TMJ'	0.9676 0.7858

(b)

Fig. 4. Results of our experiments. (a) Detection accuracies with YOLOv5 variants. (b) Results of applying EasyOCR for LP recognition with YOLOv5x

5 Conclusion

The paper reviews deep learning-based ALPR techniques. We have conducted a comparative analysis of these techniques concerning detection and acquisition accuracy. Our study has led to developing an ALPR system that recognizes license plates in English. Initially, the system identifies a vehicle in an image and subsequently detects and extracts the license plate for character recognition. We conducted experiments with various YOLOv5 models and compared their performance using Precision, Recall, and mAP@0.5 metrics. Our results reveal that the YOLOv5x model is the most efficient for license plate recognition.

Additionally, our experimental findings demonstrate that our approach achieves an mAP of $\approx 82\%$ in identifying vehicles and license plates in an image. Moreover, we achieve a confidence level of nearly 65% for recognizing license plate characters. And this also catered for different orientations of the license plates' text to a greater extent. Nonetheless, EasyOCR is ineffective in recognizing orientations and Arabic text.

The effectiveness of our approach under various weather conditions still needs to be tested. We plan to conduct future experiments to assess its efficacy in harsh environments affected by lighting and humidity conditions. We also aim to refine our proposed system to enable the recognition of Arabic license plates in the MENA region.

References

1. Ozer, M.: Automatic licence plate reader (ALPR) technology: is ALPR a smart choice in policing? Police J. **89**, 117–132 (2016)
2. Tian, Q., Guo, T., Qiao, S., Wei, Y., Fei, W.: Design of intelligent parking management system based on license plate recognition. J. Multimed. (2014)
3. Hendryli, J., Herwindiati, D.E.: Automatic license plate recognition for parking system using convolutional neural networks (2020)
4. Qi, X., Ji, Y., Li, W., Zhang, S.: Vehicle trajectory reconstruction on urban traffic network using automatic license plate recognition data. IEEE Access **9**, 49110–49120 (2021)
5. Silva, J.G., et al.: Evaluation of an LPR-based toll enforcement system on portuguese motorways. In: 2006 IEEE Intelligent Transportation Systems Conference, pp. 719–724 (2006)
6. Bochkovskiy, A., Wang, C.-Y., Liao, H.-Y.M.: YOLOv4: optimal speed and accuracy of object detection (2020)
7. Redmon, J., Divvala, S., Girshick, R., Farhadi, A.: You only look once: unified, real-time object detection. In: Proceedings of the IEEE Conference on Computer Vision and Pattern Recognition, pp. 779–788 (2016)
8. Glenn, J.: YOLOv5 by Ultralytics (2020). Accessed 17 Jan 2023
9. Punla, C., Farro, R., et al.: Are we there yet?: an analysis of the competencies of beed graduates of bpsu-dc. Int. Multidiscip. Res. J **4**, 50–59 (2022)
10. Tang, J., Wan, L., Schooling, J., Zhao, P., Chen, J., Wei, S.: Automatic number plate recognition (ANPR) in smart cities: a systematic review on technological advancements and application cases. Cities **129**, 103833 (2022)
11. Samantaray, M., Biswal, A.K., Singh, D., Samanta, D., Karuppiah, M., Joseph, N.P.: Optical character recognition (OCR) based vehicle's license plate recognition system using python and OpenCV. In: 2021 5th International Conference on Electronics, Communication and Aerospace Technology (ICECA), pp. 849–853. IEEE (2021)
12. Chen, C., Zhong, X.: Research on license plate character recognition technology based on image processing and deep learning. In: 2022 IEEE International Conference on Electrical Engineering, Big Data and Algorithms (EEBDA), pp. 1098–1102. IEEE (2022)
13. Antar, R., Alghamdi, S., Alotaibi, J., Alghamdi, M.: Automatic number plate recognition of saudi license car plates. Eng. Technol. Appl. Sci. Res. **12**(2), 8266–8272 (2022)
14. Alkawsi, G., Baashar, Y., Alkahtani, A.A., Kiong, T.S., Habeeb, D., Aliubari, A.: Arabic vehicle licence plate recognition using deep learning methods. In: 2021 11th IEEE International Conference on Control System, Computing and Engineering (ICCSCE), pp. 75–79. IEEE (2021)
15. Luvizon, D.C., Nassu, B.T., Minetto, R.: Vehicle speed estimation by license plate detection and tracking. In: 2014 IEEE International Conference on Acoustics, Speech and Signal Processing (ICASSP), pp. 6563–6567. IEEE (2014)

16. Kumari, P.L., Tharuni, R., Vasanth, I.S., Kumar, M.V.: Automatic license plate detection using KNN and convolutional neural network. In: 2022 6th International Conference on Computing Methodologies and Communication (ICCMC), pp. 1634–1639. IEEE (2022)
17. Awalgaonkar, N., Bartakke, P., Chaugule, R.: Automatic license plate recognition system using SSD. In: 2021 International Symposium of Asian Control Association on Intelligent Robotics and Industrial Automation (IRIA), pp. 394–399. IEEE (2021)
18. Laroca, R., et al.: A robust real-time automatic license plate recognition based on the yolo detector. In: 2018 International Joint Conference on Neural Networks (IJCNN), pp. 1–10. IEEE (2018)
19. Montazzolli, S., Jung, C.: Real-time Brazilian license plate detection and recognition using deep convolutional neural networks. In: 2017 30th SIBGRAPI Conference on Graphics, Patterns and Images (SIBGRAPI), pp. 55–62. IEEE (2017)
20. Riaz, W., Azeem, A., Chenqiang, G., Yuxi, Z., Khalid, W., et al.: Yolo based recognition method for automatic license plate recognition. In: 2020 IEEE International Conference on Advances in Electrical Engineering and Computer Applications (AEECA), pp. 87–90. IEEE (2020)
21. Kim, T., Kang, C., Kim, Y., Yang, S.: AI camera: real-time license plate number recognition on device. In: 2022 IEEE International Conference on Consumer Electronics (ICCE), pp. 1–6. IEEE (2022)
22. Spanu, M., et al.: Smart cities mobility monitoring through automatic license plate recognition and vehicle discrimination. In: 2021 IEEE International Symposium on Broadband Multimedia Systems and Broadcasting (BMSB), pp. 1–6. IEEE (2021)
23. Usama, M., Anwar, H., Shahid, M.M., Anwar, A., Anwar, S., Hlavacs, H.: Vehicle and license plate recognition with novel dataset for toll collection, arXiv preprint arXiv:2202.05631 (2022)
24. Tourani, A., Shahbahrami, A., Soroori, S., Khazaee, S., Suen, C.Y.: A robust deep learning approach for automatic iranian vehicle license plate detection and recognition for surveillance systems. IEEE Access **8**, 201317–201330 (2020)
25. Sahu, C.K., Pattnayak, S.B., Behera, S., Mohanty, M.R.: A comparative analysis of deep learning approach for automatic number plate recognition. In: 2020 Fourth International Conference on I-SMAC (IoT in Social, Mobile, Analytics and Cloud) (I-SMAC), pp. 932–937. IEEE (2020)
26. Elnashar, M., Hemayed, E.E., Fayek, M.B.: Automatic multistyle egyptian license plate detection and classification using deep learning. In: 2020 16th International Computer Engineering Conference (ICENCO), pp. 1–6. IEEE (2020)
27. Lee, Y.Y., Halim, Z.A., Ab Wahab, M.N.: License plate detection using convolutional neural network–back to the basic with design of experiments. IEEE Access **10**, 22577–22585 (2022)
28. Sung, J.-Y., Yu, S.-B.: Real-time automatic license plate recognition system using Yolov4. In: 2020 IEEE International Conference on Consumer Electronics-Asia (ICCE-Asia), pp. 1–3. IEEE (2020)
29. Gnanaprakash, V., Kanthimathi, N., Saranya, N.: Automatic number plate recognition using deep learning. In: IOP Conference Series: Materials Science and Engineering, vol. 1084, p. 012027. IOP Publishing (2021)
30. Minh, H.T., Mai, L., Minh, T.V.: Performance evaluation of deep learning models on embedded platform for edge AI-based real time traffic tracking and detecting applications. In: 2021 15th International Conference on Advanced Computing and Applications (ACOMP), pp. 128–135. IEEE (2021)
31. Lin, C.-J., Chuang, C.-C., Lin, H.-Y.: Edge-ai-based real-time automated license plate recognition system. Appl. Sci. **12**(3), 1445 (2022)

32. Rathi, R., Sharma, A., Baghel, N., Channe, P., Barve, S., Jain, S.: License plate detection using yolo v4
33. Batra, P., et al.: A novel memory and time-efficient alpr system based on yolov5. Sensors **22**(14), 5283 (2022)
34. Raj, S., Gupta, Y., Malhotra, R.: License plate recognition system using Yolov5 and CNN. In: 2022 8th International Conference on Advanced Computing and Communication Systems (ICACCS), vol. 1, pp. 372–377. IEEE (2022)

An Investigation into Predicting Flight Fares in India Using Machine Learning Models

Vishan Lal(✉), Paul Stynes, and Cristina Hava Muntean

National College of Ireland, Dublin 1, Ireland
vishan.tahlani15@gmail.com

Abstract. The number of passengers traveling on aeroplanes in India is increasing, and so are price changes. Seasonal and special event fluctuations occur in Indian airfares during different periods of the year. The challenge is to accurately predict flight prices. The research work presented in this paper has explored several machine learning models to predict airfares based on multiple characteristics which enhances the flight price prediction accuracy. This research proposes basic and advanced regression models that have been investigated seeking to accurately predict the price of the airline for a passenger to encourage passengers to make the flight ticket booking at the most optimal cost. The basic and advanced regression models explored in this research are Linear Regression, Decision Tree as a regressor, Random Forest as a regressor, XG Boost regressor, K-neighbours regressor, Bagging Regressor and Extra Trees regressor. The performance of these models was evaluated based on the Mean Average Error (MAE), Root Mean Square Error (RMSE), and adjusted R-Square metrics. After the evaluation of all the models that were implemented, Results analysis was discussed which showed that XG boost model achieved the best performance evaluation and the predicted prices were almost matching the actual price values of the flights.

Keywords: Basic Machine Learning models · Advanced Machine Learning models · Flight price prediction · Regression

1 Introduction

Currently, airlines use sophisticated tactics and procedures to allocate ticket pricing in a dynamic manner. These techniques take into consideration a number of financial, marketing, commercial, and societal elements that have a direct impact on the ultimate price of flight. Due to the tremendous complexity of the pricing methods used by airlines, it is very difficult for a passenger to acquire an airline ticket at the lowest price, since the price fluctuates constantly.

This research paper aims to investigate if Machine Learning models like Linear Regression, Decision Tree Regressor, Bagging Regressor, XG Boost Regressor, K neighbours Regressor and Extra Trees regressor can accurately predict the price of the airfare for a specific departure date.

M. Younas et al. (Eds.): Deep-BDB 2023, LNNS 768, pp. 106–118, 2023.
https://doi.org/10.1007/978-3-031-42317-8_9

Recently, a number of algorithms have been proposed that may offer the customer with the optimal time to acquire an airline ticket by forecasting the airfare price (Groves & Gini, 2013). The proposed solutions use complex prediction models from the sub-field of Artificial Intelligence research known as Machine Learning (ML). Specifically, (Groves & Gini, 2013) used the PLS regression model to optimize the purchase of airline tickets with an accuracy of 75.3%. Using Ripple Down Rule Learner (74.5% acc.), Logistic Regression (69.9% acc.), and Linear SVM (69.9% acc.), (Ren, 2015) investigated the accuracy of various modules like Linear Regression (77.02% accuracy), Naive Bayes (73.06% accuracy), SoftMax Regression (76.82% accuracy), and SVM (80.6% accuracy for two bins) in forecasting the price of airline tickets.

The airline industry is particularly vulnerable to the impacts of seasonality. It tends to be greater in the summer compared to the winter since vacationers and leisure travellers like to travel and take vacations during the summer when schools are out of session and the weather is nicer (Merkert & Webber, 2018). Owing to numerous events that have occurred during the year as well as due to consistent seasonality throughout the year, airlines have also consistently changed their patterns in real time. By analysing historical datasets for the variation in ticket prices, this study aims to improve the customer experience while purchasing a journey at a very low cost.

The airfare of Indian airlines varies based in various factors such as, airline stops, the airline company name, the departure city, departure time of the flight, arrival time of the flight, Destination City, Cabin Class, Duration of Flight, and the number of days left before departure. All these factors make it hard for the passenger to predict the cost of the flight and to make a booking at an optimal fare.

The traditional approaches that were being used for the flight price prediction failed to predict the prices that offer to the customers the cheapest fare. The objective of the research presented in this paper is: to analyze various ML models that can be used to predict flight prices so that passenger can make a booking at an optimal cost. This research can be applied in the airlines industry and can support customers looking to book their flight tickets at an optimal cost. The paper is structured as follows. Current related work in the investigated research area is discussed in Sect. 2. Section 3 introduces the research methodology which describes the steps taken to complete this piece of research including description of the data pre-processing mechanism and the models applied. Section 4 presents the design specification diagram that was used to implement for the flight price prediction. Section 5 presents evaluation results and discussion of the investigated models. The last section concludes the work.

2 Literature Review

Researchers have been attempting to estimate the cost of airfares in recent years by examining several factors that affect flight costs. It is difficult to determine which strategies will be more successful in predicting the flight fare. Many researchers have investigated how passengers behave while making advance purchases or how statistical and ML models can be used to determine changes of the airfares The latest research into these directions are presented next.

2.1 Empirical Approach to Determine Changes of Airfares and Customer Behavior When Purchasing Flight Tickets

As a consequence of demand shocks and distorting fluctuations in willingness to pay, airfares alter throughout time. Williams (2020) claims that a reliable flight pricing model was developed that forecasts the flight price using recent flight-level data. Using the model, it was possible to distinguish important relationships between the arrival patterns of different client categories and the lack of capacity as a consequence of demand parametric uncertainty. It has been shown that variable flight pricing boosts output by lowering the prices demanded by price-conscious customers who buy in advance. Furthermore, it ensures seats for those latecomers with the greatest willingness to pay (such as business travellers), who are subsequently charged premium fees. Furthermore, it was shown that dynamic airline pricing increases total welfare in comparison to a more constrained pricing structure.

2.2 Statistical Approaches for Determining Changes in the Airfare

Using artificial intelligence models, (Groves & Gini, 2013) coupled the PLSR (Partial Least Square Regression) model to get the largest presentation for the least cost of purchasing airplane tickets, with an accuracy of 75.3%. Ren (2015), took into consideration the application of models such as linear regression (with a precision of 77.06%), naive bayes (with a precision of 73.06%), SoftMax regression (with a precision of 76.84%), and support vector machine (with a precision of 80.6%) to estimate the cost of airline tickets.

Groves & Gini (2013) used the Partial Least Square Regression (PLSR) technique to construct a model that was capable of predicting the ideal time to buy airline tickets. The information was obtained from the most popular online travel booking services between the 22nd of February 2011 and the 23rd of June 2011. Additional data were also obtained, and these data are now being put to use in order to validate the comparisons of the final model's performances.

Dutta & Santra (2017) looked at the domestic airline market's dynamic price dispersion. The average power divergence statistic (PDS) and average airfare movement were computed for each route. The impact of Average PDS was assessed based on a few selected route characteristics and market structure parameters. The research performed by (Dutta & Santra, 2017) demonstrates that prices increase as the departure date draws near if both full service airlines and low-cost carriers are present in the same market. Revenue management and dynamic pricing are often used in the domestic aviation industry in India. Additionally, it suggests that a route's characteristics may affect how and where airfares are distributed.

To investigate the price dispersion of U.S. airlines, (Merkert & Webber, 2018) suggest a model of rational airline seasonal behaviour. The model was developed, calibrated, and tested using airlines from the Asia-Pacific region, Europe, North America, and South America. The statistics show that seat factors and airline costs fluctuate seasonally. Contrary to logic, seasonal fluctuation in average airfare is greater than that in seat factor. The high-to-low average price ratio need to be less than the high-to-low seat factor ratio. Like airfare, seat factor should be seasonal. Pricing has long been the main topic of competitive strategy and revenue management literature.

Price factors for airline tickets include flight distance, purchasing window, fuel cost, and others. Each carrier has unique algorithms and price-setting standards. These principles may be inferred using recent developments in AI and ML, which can also mimic price movement. An innovative application is provided by Wang et al. (2019) and is based on the ACS and DB1B public air transportation datasets (T-100). The suggested approach combines the two databases with macroeconomic data, applies machine learning methods, and uses origin and destination pairings to estimate the quarterly average ticket price. With an adjusted R squared of 0.869, the method correctly predicts the testing dataset.

The statistical approaches used in the above presented works produced better patterns in detecting the flight price variation along the days left before departure in the US, but the models that were considered as part of the above presented works took time series models and took only days to departure, day of the week features and the derived time series features from these 2 features into consideration mainly to determine the flight price pattern movement, the above presented works doesn't take into account, the other important features like the flight distance, flight departure time, flight source city, flight destination city, flight arrival time and flight duration, which are also responsible to create the flight price movement pattern before the departure date. This research takes various important features into account to determine the flight price movement pattern and generates more accurate model to predict the flight price movement pattern.

2.3 Supervised Machine Learning for Determining the Changes in the Airfares

The research (Wozny, 2022) examines machine learning predictions for hypothetical scenarios involving air travel. As an example, the impact of COVID-19 policy changes on 2020 airfare was predicted. Studying airfares is crucial since air travel is essential for mobility. Results indicate that approach influences the accuracy of predictions. OLS Had a significant prediction error for counterfactual scenarios, while ML outperformed it. prior to 2020, forecasts made by ML models were precise. The RMSE of airfares was 5 between 2012 and 2019, fluctuating between USD 242 and 134. The discrepancy between actual and anticipated airfare in 2020 was 25 USD, which was three times more than in previous years. 2020 airfares would be higher without COVID-19.

Uncertain decision-making is a challenge for modern computer-aided approaches and applications. Here, Bayesian prediction techniques are helpful. Boruah et al. (2019) used the Kalman filter, a Bayesian estimate method, to anticipate flight prices. The linear Kalman Filter model is used in this technique. Based on previous fares, this model predicts future flight prices. The linear model receives observed data as a matrix, then computes an expected ticket price for an upcoming flight. It is challenging to forecast airline ticket prices and demand since both internal and external factors are subject to rapid change. To assist consumers or airlines in estimating demand, (Abdella et al., 2019) have created alternate ticket price/demand prediction algorithms. Customer-side and airline-side prediction models looked into it. Both types of models depend heavily on historical ticket prices, purchase dates, and departure dates. Search engine results and information from social media are not regarded as external components.

The airline Optimal Ticket Purchasing Decision-Support Service (OTPS) that (Xu & Cao, 2017) suggests identifies the optimal window of opportunity prior to departure. The DPLP technique, which considers variations in airline fare, is the foundation of OTPS. To increase the generality and reliability of the OTPS, parameter values are route-specific and often modified. The real-world ticket price dataset for many routes is utilized for in-depth analyses. Experiments show that OTPS outperforms other state-of-the-art solutions.

The above presented works which were implemented using supervised Machine Learning techniques to predict the airfares had emphasis on different features which didn't have much impact in determining the airfares in India, also the above presented research works didn't take various optimal ways to build the model into account and generate more accurate prediction metrics. This research takes most important features required to predict the flight prices for the Indian airlines market and also the model implementation is optimally performed to predict the flight prices in India.

3 Research Methodology

The research methodology consists of five **steps namely** Application Understanding, Data Gathering, Data pre-processing and transformation, Data modelling and conversion, Evaluation and results (as shown in Fig. 1).

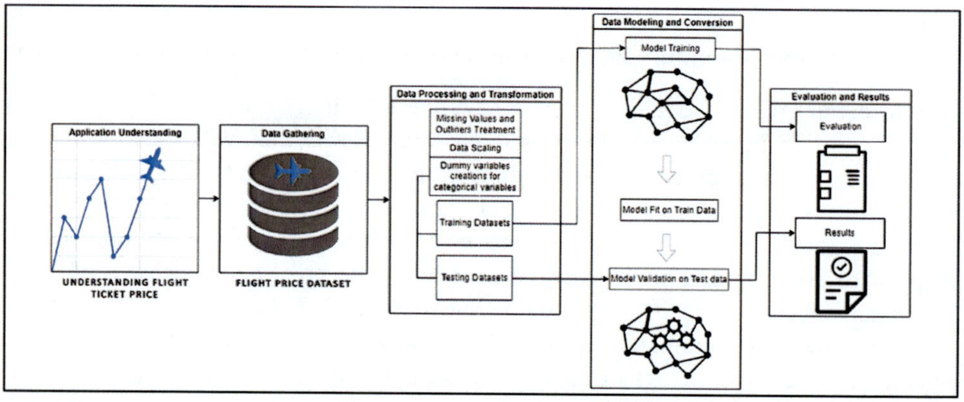

Fig. 1. Research Methodology

The first step *Application Understanding* aims to explain that the Variability in the airfares can be analysed effectively by using various Machine Learning techniques. The airfares movement are being judged manually based on the human sentiments based on their experiences, which lacks to take into account the other factors which affect the variability in the airfares and also this manual process takes a lot of time to identify the right price for the flight for a specific departure date for a customer. Recognizing that traditional techniques might, at times, be inaccurate and realizing that using modern ways will result in findings that are more accurate and produced more quickly. This research can enhance in automating the customer's experience to make the booking of the flight at the most optimal cost. The passengers will be able to make flight booking at an optimal cost by getting the predictions and also determine the number of days in advance for a passenger to make the flight booking, by predicting the flight ticket price pattern with respect to number of days left for the departure.

The second step *Data Gathering* involves the gathering of dataset to predict the flight price prediction is being taken from the Kaggle[1] which is a public and open source repository. The Dataset captures the flight level data over the period of 11th February 2022 to 31st March 2022, i.e. for 50 days**.** The dataset collected consists of the flight level data for the top 6 major metro cities of India and also it captures the data for all the major airline companies of India. Totally, 300261 datapoints and 11 features have been taken into consideration in the dataset. The features considered for the dataset consists of the following features, the name of the airline company, City from which the flight takes off, Departure Time of the flight, Airline Stops, Arrival Time of the flight, Destination City, Cabin Class, Duration of Flight, Days left before departure and the target variable for this dataset is considered as the price of the flight. The datasets that are taken into consideration considers the flight level data for 6 major cities of India which is Delhi, Mumbai, Chennai, Hyderabad, Kolkata and Bangalore, and also the flight level dataset captures the data from 6 major airline service providers which are Vistara, Air India, Go First, Indigo, Air Asia and Spice Jet, among all of these airline companies Vistara and Air India are known to be premium airline service providers and Go First, Indigo, Air Asia and Spice Jet are marked as the low cost airline carriers. The below correlation matrix (as shown in Fig. 2) was plotted to showcase the feature importance of each feature has on determining the flight price. It can be inferred from the correlation diagram that class of airline and airline brand plays a significant role in driving the prices of then airline tickets. The duration of the flight also has a moderately significant impact in driving the prices of the airline tickets.

On the contrary, the source and the destination city had the least significant impact on the pricing of the airline tickets.

The third step *Data pre-processing and transformation* involves to pre-process and transform the flight level dataset which is a structured dataset that requires to be undergone through various steps of the processing and transformation, the following steps were performed for the pre-processing and transformation of dataset, Initially the dataset for flight_details.csv was read into a data frame and then, the Unnamed column was dropped from the flight_details.csv dataset. After that, the dataset was diagnosed for the null values and no null values were found in any of the columns of the dataset. Then, the distribution of the numerical variables were seen by box plots to check for the outlier values in the datasets, the log scaling was applied on the features having outliers value and then the z-score method was applied to detect and remove the remaining outliers. After that some of the steps were performed to transform the dataset so that it can accurately fit into various machine learning models. Firstly, the dummy variables for all the categorical data in the pre-processed dataset is generated using the one hot encoding technique. The previous research has shown that the accuracy of a machine learning models increase when one-hot encoding for categorical variables are being applied as it

1 https://www.kaggle.com/datasets/shubhambathwal/flight-price-prediction?select=Clean_Dataset.csv.

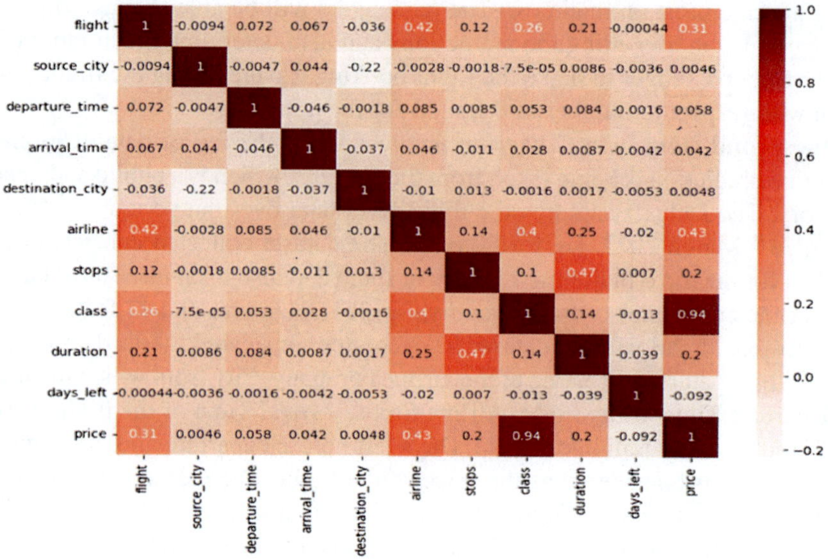

Fig. 2. Correlation matrix for all the features of flight dataset

allows the representation of categorical data to be more expressive. Then, the correlation matrix is plotted for all the dummy variables and the original feature with the price of the airline to check the impact of all the variables on the price of the airline. After this, split is made between the training and testing sets in the ratio of 70:30 (i.e. train: 70% and test: 30%) to apply various machine learning models on the pre-processed dataset and after that the target variable i.e. the price of the airline is being separated with the features set. The Min max scaling transformation was applied to the train and the test sets to have a better fit into the machine learning models.

The fourth step *Data modelling and conversion* involves to predict the continuous target variable i.e. the flight price, various ML models which are based on regression techniques are applied on the processed dataset which includes basic and advanced ML models. Regression analysis is a statistical method for connecting a dependent variable to one or more independent (explanatory) variables. A regression model may demonstrate whether variations in the dependent variable are related to variations in one or more explanatory variables. The regression models that were used to predict the prices of the flight tickets were, multiple linear regression, decision tree regressor, K-neighbours regressor, extra trees regressor, XGBoost regressor and bagging regressor.

The fifth step *Evaluation* involves to evaluate various metrics for the machine learning regression models implemented to give an accurate prediction of the flight price and compare these metrics to get the best model to accurately predict the flight prices. These evaluation metrics are RMSE (Root Mean Squared Error), MSE (Mean Squared Error), MAE (Mean Absolute Error), Adjusted R-Square. Graphical analysis on the performance of all the models were obtained and the model with lowest error terms was taken into consideration to fit the test data and get the flight price prediction on the testing data. The detailed description and formulas for the evaluation metrics used to measure the prediction of airline ticket prices can be found here.[2]

4 Design Specifications

In order to successfully predict the prices of the airfares it becomes important to comprehend the description of the design approach that I will be following to predict the cost of the air fares. The airfare prediction starts by collecting the Airlines dataset from Kaggle which is a public open source repository and then followed by the pre-processing various features of the data and after getting the pre-processed data, the data is split into train and testing sets and then various regression based models are applied on the training data, followed by the testing of the model on the test data and finally a predicted output of flight price is generated for the test data and the model is evaluated. The below diagram (as shown in Fig. 3) explains about the same.

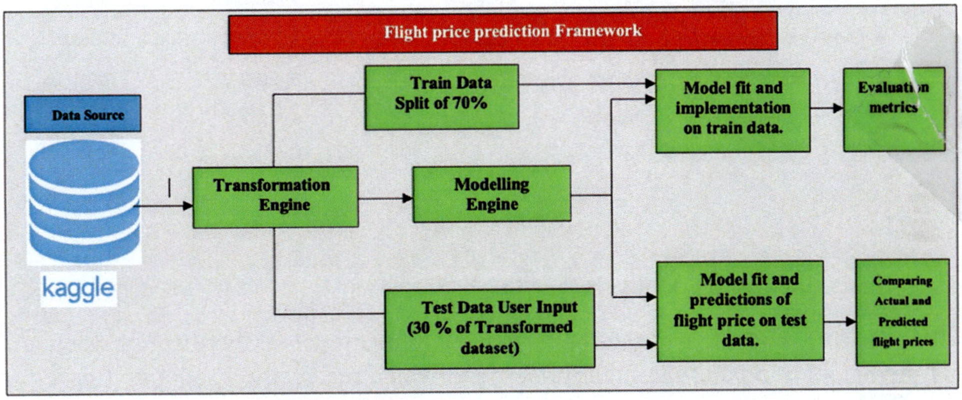

Fig. 3. Architecture diagram for Airfare prediction

5 Evaluation Results and Discussion

The objectives of evaluation stage of this research were to predict the cost of flight tickets which are least deviated from the actual flight prices. Evaluation metrics such as RMSE, MSE, MAE and Adjusted R-Square were calculated and compared for the ML modules considered in this research. The models are used for predicting the price of the flight on the training data and values of the metrics are shown in Table 1. The best model i.e., the model which has the highest adjusted R-Squared and the lowest RMSE value was then applied on the test data to predict the flight ticket price and it was compared with the actual flight price of the test data. The scikit learn package of Python language

[2] https://medium.com/analytics-vidhya/mae-mse-rmse-coefficient-of-determination-adjusted-R-squared-which-metric-is-better-cd0326a5697e#:~:text=The%20lower%20value%20of%20MAE,variability%20in%20the%20dependent%20variable.

has been used to implement these Machine learning (ML) models and parameters were tuned for all the implemented models. Further on, the scikit learn packages has been used to calculate the metrics for all the implemented models.

An analysis of the results presented in table is classified into Ensemble model analysis (Advanced Regression models) and Basic Regression models which are presented in Sects. 5.1 and 5.2 respectively.

Table 1. Comparative results for ML Models results and metrics for computing flight prices from training data

	Model Name	Adj R-Squared	Mean Absolute Error	Root Mean Squared Error_RMSE	Mean Squared Error_MSE
0	XGBoost Regressor	0.984567	1131.313984	2818.111048	7941750.0
1	K-Neighbours regressor	0.982006	1209.984116	3043.043307	9260113.0
2	Extra Trees Regressor	0.978741	1819.477614	3307.636674	10940460.0
3	Decision Tree Regressor	0.973809	1247.902298	3671.225568	13477900.0
4	Gradient Boosting Regressor	0.970023	1881.169731	3927.696573	15426800.0
5	Bagging Regressor	0.906950	4572.481049	6919.907473	47885120.0
6	Linear regression	0.906949	4571.189881	6919.960632	77885860.0

5.1 Ensemble Model Analysis

The **boosting models** performed better than **bagging models** as there were fewer number of predictors to predict the flight prices, and which means that model will have more bias and low variance and hence, to treat model bias and reduce the model underfitting boosting models are the most optimal and efficient models.

The **XG boost** regressor produced the best set of results in training data with adjusted R-Square value of 0.9845, MAE as 1111.31, RMSE as 2818.11 and MSE as 7.941×10^6 which means that the predicted prices vary by the factor of 0.9845 from the original prices and the average deviation for the predicted prices are deviated by the value of 1111.31 INR (Indian National Rupees). This result metrics indicated an improved prediction model by an increase 17% in adjusted R-Square value as compared to the research conducted by (Wang et al., 2019), in which Random forest was proven to be the best model, since the model used by (Wang et al., 2019) had higher bias and lower variance and hence more emphasis should have been given to improve the boosting models, rather than improving the bagging model to get the better accuracy results. The XGboost model applied in this research performs better than other boosting models as it takes L1 and L2 regularization into account to reduce the model bias and also by supplying optimal and efficient parameters during the model building, which in turn helped to reduce the model underfit more efficiently to predict flight prices. The process was carried over by boosting the model in the subsequent boosting stages and to reduce the model prediction bias subsequently in the model boosting stages and avoid the model underfitting, this was achievable by adjusting the number of boosting stages required to accurately predict the flight prices this was done by adding the n_estimators parameter while building the model, Also, the efficient number of nodes in the decision trees were set during each of the boosting stages by configuring the max_depth parameter in the model. The model was also tuned in a way so that it didn't get the model overfitting issue while predicting the flight price this was possible by adding efficient learning_rate in the model parameter.

The **Extra Trees regressor, Gradient Boosting regressor and the Bagging regressor** models were also implemented which gave closer results to predict the flight prices with their adjusted R-square values as 0.9787, 0.9572 and 0.9069 respectively. The normal **Boosting** models performed lower than the **XG boost regressor** as these models didn't take into account of the L1 and L2 regularization which is required to reduce the model underfit even more than the usual boosting stages which makes the bias even lower than the regular boosting algorithms.

The **Bagging regressor** couldn't produce better results because the model had more bias and low variance as there were few predictors, and bagging reduces the variance, and lowers the model overfitting issue, but since the model used to predict the flight prices already has lower variance so bagging didn't perform as good as the boosting models, since in bagging, each model receives an equal weight whereas in boosting, models are weighed based on their performance from previous iterations. Also, Models are built independently in Bagging whereas new models are affected by a previously built model's performance in Boosting. During the bagging model build, training data subsets are drawn randomly with a replacement for the training dataset. In Boosting, every new subset comprises the elements that were misclassified by previous models. Hence, Boosting models had better model performance to predict the flight prices than the bagging models.

5.2 Basic Machine Learning Model Results

The **K-neighbours** model was the best performing basic machine learning model. The performance was slightly low achieving an adjusted R-Square of 0.9820, MAE of 1209.98, RMSE of 3043.04 and MSE of 9.2 $x10^6$, which means that the predicted prices vary by the factor of 0.9820 from the original prices. The average deviation for the predicted prices are deviated by the value of 1208.98 INR.

The **Decision Tree and Linear regressor** model were the second and third best performing basic machine learning model respectively, with the adjusted R-Square value of 0.9738 and 0.9069 respectively, MAE as 1247.90, 4571.18 respectively and RMSE as 3307.63, 6919.96 respectively. Therefore, it can be inferred that the predicted and the actual values are very close and the average deviation for the predicted prices for **Decision Tree and Linear regressor** are deviated by the value of 1247.90 INR, 4571.18 INR respectively.

Since the **XG boost** model had the best evaluation metric values in predicting the flight price, the model was fitted on the test data to get the flight price predictions on the test data. Figure 4 (as shown in Fig. 4) showcases the prediction of flight prices on the test data and plotted against the actual values of the flight prices. The test data was seen to fit well on the XG Boost model and the predicted prices were almost matching the actual price values of the flights.

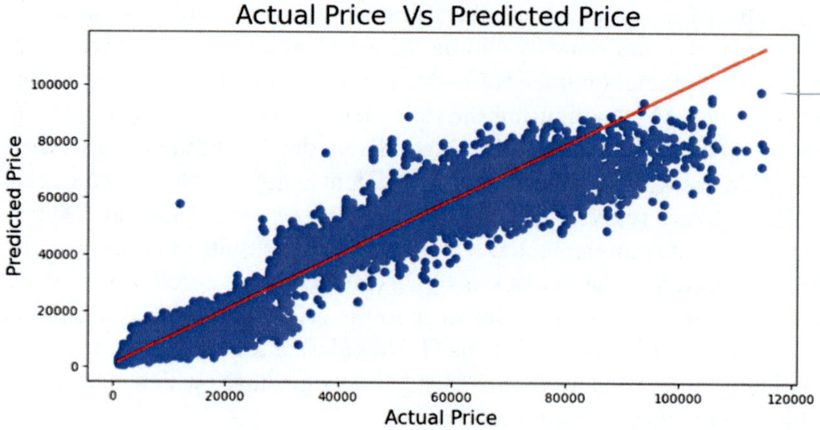

Fig. 4. Actual vs the predicted prices of the flight

Furthermore, the actual and the predicted prices found on the test data of the flights were plotted with respect to the number of days left for the departure and the price of the flight (as shown in Fig. 5). The predicted price before the number of days left for departure almost matched the predicted values, which were found by fitting XGBoost model on the test data which took the following features such as, flight departure time, flight source city, flight destination city, flight arrival time and flight duration, flight seating class, flight stops and days left before departure into consideration. This prediction pattern for flight price movement was relatively better than the existing work presented by Groves & Gini (2013) in which, the statistical approaches were mainly used. The models that were considered by Groves & Gini (2013) were mainly time series models, that took only days to departure, day of the week as features and then derived time series features from these 2 features into consideration mainly to determine the flight price pattern movement, the model used by Groves & Gini (2013) doesn't take into account, the other important features which are also responsible to create the flight price movement pattern before the departure date, these features were considered as part of this research.

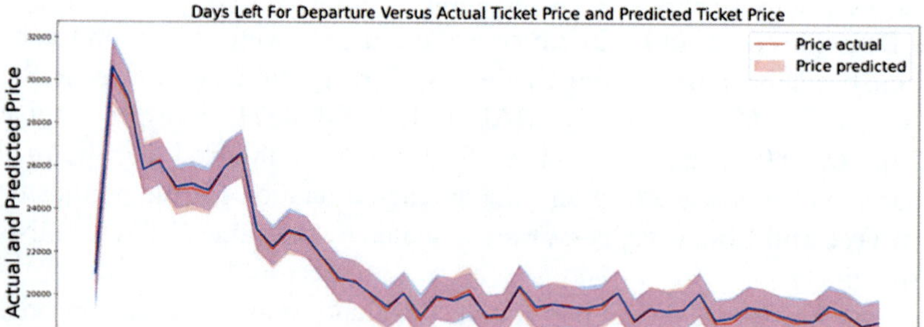

Fig. 5. Days Left For Departure Versus Actual Ticket Price and Predicted Ticket Price

6 Conclusion and Future Work

The flight price prediction is an important aspect to be researched as most passengers who plan to book flights are looking for an optimal cost. The literature review has shown various research directions and approaches related to flight price prediction. This research paper has investigated six machine learning models to predict airfares based on multiple characteristics. The models were evaluated using important metrics such as RMSE, MSE, MAE and Adjusted R-Square. Based on models performance analysis the best model was tested on the test dataset to predict the price of flight XG boost model had the best performance evaluation and the predicted prices were almost matching the actual price values of the flights. With an adjusted R-squared value of 0.9845, when using XG boost model, the prediction was better than one achieved in research conducted by Wang et al. (2019) that produced the best adjusted R-squared value of 0.858 when using the Random forest model.

There are a few limitations that can be taken care as part of the future research work such as the dataset contains record values for 50 days in a year, which limits the research to generate patterns of the price movement of airfares in a limited amount of time over the year; and the seasonality in airfares can't be interpreted well for the entire year. The dataset considered for this research consists of airlines data for 6 major cities of India. Having flights related information in the dataset for more cities from India, prediction of airfares for other Indian cities could be performed.

References

Abdella, J.A., Zaki, N., Shuaib, K.: Automatic detection of airline ticket price and demand: a review. In: Proceedings of the 2018 13th International Conference on Innovations in Information Technology, IIT 2018 (2019). https://doi.org/10.1109/INNOVATIONS.2018.8606022

Boruah, A., Baruah, K., Das, B., Das, M.J., Gohain, N.B.: A Bayesian approach for flight fare prediction based on kalman filter. In: Panigrahi, C.R., Pujari, A.K., Misra, S., Pati, B., Li, K.-C. (eds.) Progress in Advanced Computing and Intelligent Engineering. AISC, vol. 714, pp. 191–203. Springer, Singapore (2019). https://doi.org/10.1007/978-981-13-0224-4_18

Dutta, G., Santra, S.: An empirical study of price movements in the airline industry in the Indian market with power divergence statistics. J. Rev. Pricing Manag. **16**(2) (2017). https://doi.org/10.1057/rpm.2016.12

Groves, W., Gini, M.: An agent for optimizing airline ticket purchasing. In: 12th International Conference on Autonomous Agents and Multiagent Systems 2013, AAMAS 2013, vol. 2 (2013)

Merkert, R., Webber, T.: How to manage seasonality in service industries – the case of price and seat factor management in airlines. J. Air Transp. Manag., **72** (2018). https://doi.org/10.1016/j.jairtraman.2018.07.005

Wang, T., et al.: A framework for airfare price prediction: a machine learning approach. In: Proceedings of the 2019 IEEE 20th International Conference on Information Reuse and Integration for Data Science, IRI 2019 (2019). https://doi.org/10.1109/IRI.2019.00041

Williams, K.: Dynamic airline pricing and seat availability. SSRN Electron. J. (2020). https://doi.org/10.2139/ssrn.3611696

Wozny, F.: The impact of COVID-19 on airfares—a machine learning counterfactual analysis. Econometrics, **10**(1) (2022). https://doi.org/10.3390/econometrics10010008

Xu, Y., Cao, J.: OTPS: a decision support service for optimal airfare ticket purchase. In: Proceedings of the 2017 IEEE International Conference on Big Data, Big Data 2017, 2018-January (2017). https://doi.org/10.1109/BigData.2017.8258068

Securing Internet of Things (IoT) Devices Through Distributed Ledger Technologies (DLTs) and World Wide Web Consortium (W3C) Standards

Sthembile Mthethwa(✉)

Council for Scientific and Industrial Research (CSIR), Information and Cyber Security Centre (ICSC), Pretoria, South Africa
smthethwa@csir.co.za

Abstract. The tremendous growth of the Internet of Things (IoT) connected devices calls for a way for manufacturers to standardise the newly created devices for security purposes. As these devices are getting smaller and smaller so does the computational power in them. The resource constraints introduced by the devices eliminate some of the cryptographic features that can be performed by these devices. Thus, developers are looking for a secure method for automating processes and exchanging information in real-time. With that, new technologies like distributed ledger technologies (DLTs) introduce a promising solution for enabling large-scale IoT applications in a decentralized and autonomous manner. Therefore, this work aims to investigate the integration of DLTs and IoT to create a safe and secure environment for IoT devices whilst being aware of the constraints brought by these devices. In this paper, a proposed solution is presented that adopts emerging standards introduced by the World Wide Web Consortium (W3C), to ensure an internationally recognisable system and platform.

Keywords: Internet of Things (IoT) · Distributed Ledger Technologies (DLTs) · World Wide Web Consortium (W3C) · IoT devices

1 Introduction

The Internet of Things (IoT) is a network of devices consisting of sensors that can detect data from the physical environment and can communicate and interact with its surroundings [1, 2]. IoT turned conventional objects into smarter ones. There has been a tremendous increase in the adoption of IoT in various sectors. It is projected that there will be more than 75 billion IoT-connected devices in use by 2025 [3]. The continual use of IoT devices span to various sectors i.e., health (for medical equipment), governments or municipalities (smart devices to monitor energy use, and water and air quality), agricultural sector (monitor crop conditions—light levels, humidity, soil moisture, crop health, and more—and then automate irrigation and other processes accordingly) and engineering or operational technology (OT – which is part of IoT devices) (identify,

M. Younas et al. (Eds.): Deep-BDB 2023, LNNS 768, pp. 119–128, 2023.
https://doi.org/10.1007/978-3-031-42317-8_10

monitor, and control physical devices, processes, and events). Among these applications are many safety-critical tasks, where a malfunction or malicious tampering with devices can impact human health, may have a large financial impact, or tampers with privacy. Their increasing spread and importance also increase the interest attackers are developing for IoT-devices. [4, 5].

IoT devices collects data which is usually stored in a centralised cloud storage for analysis and processing by various applications. Consequently, the data becomes vulnerable to various forms of attacks and compromises its security [6].

With the advancements in IoT devices, whereby, the size of devices is significantly reduced, consumes less energy and reduced hardware cost which enables them to be integrated into everyday objects. Another issue is introduced due to the heterogeneity of IoT devices, whereby the architecture varies per device. These devices are manufactured by various companies with different specifications. It is worth noting that, with all the advancements in technology, IoT devices are still susceptible to challenges, such as:

- Constrained resources - IoT devices have constrained resources with respect to central processing unit (CPU) processing power, read-only memory (ROM), Random Access Memory (RAM), and battery life. However, these devices still have the capability of providing their intended functionalities such as collecting and transmitting data, across the Internet for storage and analysis [7].
- Privacy - these are issues relating to the collection, storage, use, and sharing of personal information. The vast amount of data generated by IoT devices raises privacy concerns, as personal information could be collected and used without consent.
- IoT device security – there are currently no proper measures to protect IoT devices from cyber-attacks, hacking, data theft, and unauthorized access. This can be due to various reason i.e., outdated software, weak passwords, unpatched vulnerabilities, lack of encryption, etc. Therefore, to ensure the security and privacy of sensitive information stored on these devices, it is essential to implement strong security measures.
- Interoperability – is the ability of different systems, devices, or components to work together seamlessly and exchange data effectively. Therefore, ensuring that different IoT devices can work together seamlessly, and exchange data effectively is essential.
- Identity management - which ensure that the right users have the right privileges in terms of accessing devices, data and applications, and monitoring their usage. Identity management in IoT is currently performed by exchanging identifying information between devices for first time connection. This process is susceptible to eavesdropping which can lead to man-in-the-middle attack [8]. Hence, there is a need for a unique identity management solution [9].
- Lack of standardization - the absence of agreed-upon specifications or protocols in a particular field or industry which can result in different systems, products, or processes being incompatible with each other (thus, leading to inefficiency, and decreased interoperability). For example, in the context of IoT, this can cause difficulties in communication and data exchange between various IoT devices and systems. Establishing standards and protocols can help overcome this and ensure uniformity and compatibility. There is a lack of standardization in IoT devices, making it difficult to secure them consistently.

With the rapid increase of connected devices and services, it is vital for IoT device manufactures, and application developers to invest in a secure methods of automating processes and exchanging information in real-time [10]. Hence, to ensure secure IoT devices, new technologies like the distributed ledger technologies (DLTs) can be utilised which appears to be a promising solution for enabling large-scale IoT applications in a decentralized and autonomous manner. The first ever implementation of DLTs was experienced through the inception of Bitcoin, a peer-to-peer version of electronic cash that allows online payments to be sent directly from one party to another without going through a financial institution [11]. As an open, trust-less, transparent, immutable and distributed ledger, a DLT can record transactions among IoT devices in a verifiable and permanent way [10].

Over the years, the use of DLTs has shifted from the mere use of transacting cryptocurrencies but the main characteristics of DLTs have evolved and now includes various sectors like supply chain, manufacturing, etc., Below are some of the prominent features provided by DLTs that might be of interest in the field of information security and IoT:

- Decentralisation – no central entity controlling the network which can introduce a single point of failure. Instead, the network is made up of various nodes that work together to verify and validate transactions.
- Anonymity – the identity of participants is either anonymous or pseudonymous, thus improving privacy.
- Security – cryptography is used to sign data in order to prove that a transaction was approved by the owner thus, ensures that untrusted parties can communicate securely.
- Immutability – DLTs are permanent and unalterable network. Once a transaction is recorded, it cannot be modified or deleted. Thus, providing a high degree of security and trust. To guarantee data confidentiality and integrity, this quality is essential.
- Distributed - all network participants have a copy of the ledger for complete transparency. Information is not processed through a central server but is transmitted and verified by nodes in parts of the network. Allowing for a peer-based network that can self-check.
- Traceability – enables any authorized entity to authenticate a transaction's history, if there is a need to prove the origin of a transaction.
- Transparency – all participants in the network can access information in the network according to the permission they have been granted. For example, in a public network anyone can access and view all the transactions on the network.
- Consensus – a decision-making algorithm for the group of nodes active on the network to reach an agreement and for the smooth functioning of the system. In DLTs nodes do not trust each other but they can trust the algorithm that runs at the core of the network to make decisions.
- Auditability – since data in the network is reliable, accurate, verifiable, and cannot be altered; it allows the network to create an audit trail.

Considering these features, the use of DLTs can allow IoT devices to communicate among themselves and make decisions automatically. However, for a successful integration of DLTs with IoT, the following must be considered:

- Limitations of IoT devices - most DLT implementations rely on consensus algorithms (the process by which nodes in a network agree on a common state of the ledger) i.e., mining. This requires huge resource capacity which becomes computationally intensive for IoT devices. Majority of IoT devices are resource constrained (computational capacity, power, and storage), hence, the use of mining might not work as it utilizes a lot of resources (it is computationally intensive).

- Latency –ledgers maintain a history of all transactions which for example in a blockchain implementation, transactions are organised in batches known as blocks. However, not all DLTs employ blocks. Therefore, computational latency is important, which refers to the average time until blocks (or transactions) are added to the DLT so that the likelihood of tampering of previously added blocks or transactions is below a certain threshold. The approach of using blocks is time consuming which is not suitable for over-the-top IoT applications as they require low latency.
- Scalability – currently, DLTs do not scale very well as the number of nodes increases in the network, which will pose a challenge as IoT networks are increasing at a high speed and the number increases tremendously.
- Centralisation - current approaches in IoT implementations are largely centralised, which raises several security concerns i.e., single point of failure, trust, and privacy.
- Overhead traffic - which may be undesirable for certain bandwidth limited IoT devices.

In this paper a DLT-based architecture for IoT that delivers lightweight, standardised, decentralised, secure, and scalable platform is proposed. It aims to retain the benefits of DLTs while overcoming the aforementioned challenges of IoT devices. This architecture aims to be application–agnostic and well suited to diverse IoT use cases. The proposed solution aims to be in line with emerging standards aimed at standardising the process of identification.

The remainder of the paper is structured as follows: In Sect. 2, an overview of IoT and DLTs is provided as well as highlighting IoT challenges. In Sect. 3, current DLTs-based applications and services for IoT are provided. In Sect. 4, we present the proposed architecture and finally, conclusion and future work is provided in Sect. 5.

2 Overview of IoT and DLTs

2.1 Overview of IoT

The interconnectedness of IoT enables real-time collection and monitoring of various types of data about i.e., properties, individuals etc. [12]. Sensors and actuators as well as heterogeneity and decentralisation are the key features of IoT [13].

The interconnectedness combine with lightweight nature of IoT devices, makes them susceptible to attacks. As IoT devices communicate with each other, a crucial aspect of IoT security is the ability to trust in data received from another device that is part of the network during the communication process [4] which is typical for wireless sensor networks (WSN). According to [12], WSNs "are ad hoc networks that are considered the major building blocks for IoT devices. They are used for gathering data from their surrounding and delivering them to users and for accessing connected IoT devices remotely". The communication between the Internet and the sensor nodes should satisfy secrecy, trustworthiness, verification, and non-revocation [12].

The main challenge is that IoT devices are constrained by energy, memory, and processing power. Another challenge is that they experience great data losses due to node impersonation. For example, an attacker gaining unauthorized access to an application and taking control of it. Therefore, it is infeasible for traditional security mechanisms, e.g., encryption of memory and data transmission, requiring strong computational power, to be applied on IoT devices as it contradicts its light-weight nature [4].

The number of devices in the IoT network are expected to increase tremendously in the future. Therefore, improving IoT device security is crucial.

2.2 Overview of DLTs

DLTs provides a universal data structure which combines a group of previously untrusted nodes in a distributed environment thus eliminating the need of a centralized third party to oversee everything that happens in the network whilst proving immutability [14]. The first ever successful implementation of DLT was blockchain, which was introduced by bitcoin the first decentralised digital currency [11]. It is noteworthy that, all blockchains are distributed ledgers, but not all distributed ledgers are blockchains [14]. Subsequently, other dlts spanning from cryptocurrencies have been introduced over the years.

- Blockchain is a shared, decentralized, and immutable ledger of timestamped series of transactions [15]. Blockchain is based on a peer-to-peer topology as well as cryptography. To store a transaction in the ledger, most participating nodes in the blockchain network should agree and record their consent. Each new block includes a link to the prior block in the chain through cryptography. Each block encompasses a timestamp and hash function to the previous block. The participating nodes of the blockchain network place their trust in the integrity and security features of the consensus mechanism.
- Directed Acyclic Graph (DAG) is a directed graph data structure that uses a topological ordering [14]. In DAG, transactions are linked to one but possibly more transactions. However, links are specifically directed – pointing from prior transactions to newer ones in accordance with topological order. DAGs are acyclic; thus, it is a non-circular structure (whereby loops are not permitted) [15]. DAG does not add blocks sequentially (resulting to higher throughput), does not require proof of work from miners, zero transaction fee, provides higher level of scalability and is partition tolerant, which allows a portion of the network to split off the main network for a period and continue to run without the Internet connectivity [14]. There are several applications that use DAG such as IOTA (built specifically for IoT), Byteball, Hashgraph, etc.
- Hybrid DLTs are a combination of DAGs and blockchain technologies and one example is the Tempo ledger. It is an essential part of Radix, a DLT platform that works efficiently with IoT [14]. Tempo uses partitions of the ledger to accomplish the appropriate ordering of actions that occur in the whole network. The Tempo ledger comprises of three main elements; a networked cluster of nodes, a global ledger database that is distributed across the nodes, and an algorithm for generating a cryptographically secure record of temporally ordered events.

3 DLT-Based Applications and Services for IoT

The topic of integrating DLTs with IoT has gained a lot of traction over the years. This is fuelled by the improvements introduced by DLTs [16]. However, most research around this topic have mainly been focused on conducting surveys and theoretical work [17, 18]. This has left a research gap for practical implementations of DLTs in the information security space; thus, presenting an opportunity for this study to make a plausible attempt to bridge the gap.

The first implementation occurred when IBM in partnership with Samsung designed a platform called ADEPT (Autonomous Decentralized Peer-To-Peer Telemetry) which utilises the design of bitcoin to construct a distributed network of devices [19]. In 2017, Slock.it was introduced which aims to provide the transparency and auditability features to the IoT objects by integrating blockchain with IoT [20]. It resolves the problem of connecting a device to the blockchain and improves the essential features for non-blockchain designers working on IoT systems by providing an interoperable and decentralised platform [15].

Chain of Things which provides an integrated blockchain and IoT hardware solution to solve IoT challenges regarding identity, security, and interoperability was introduced [21]. A survey of types of distributed consensus or trust protocols from an IoT perspective, has been conducted by [5] which provides trade-offs and recommendations. This would be vital for using DLTs for purposes of securing IoT devices.

To the best of our knowledge, SmartDID is the only distributed-identity management system for IoT that implements DIDs standards to preserve privacy [22]. Hence, the gap for ensuring IoT devices security using DLTs and standards.

4 Proposed Architecture

This section provides details about the proposed architecture for the integration of these two technologies and how we envision it working. The planned prototype would i.e., visualise and demonstrates the enrolment of devices, how IoT devices share data or communicate, assigning identities to the devices etc., whilst considering storage throughput in the order to cater for multiple of transactions per second between IoT devices.

The major contribution of this research is to propose a solution aimed at ensuring security is maintained even for resource constrained IoT devices. To achieve this, a DLT has been employed to benefit from some of its prominent features discussed in Sect. 1 and eminently solve some of the challenges outlined for IoT devices i.e., security, centralisation of data, privacy etc.

To solve the issue of standardisation in IoT devices, this paper aims to adopt standards introduced by the World Wide Web Consortium (W3C) which includes the use of Verifiable Credentials (VCs) and Distributed Identifiers (DIDs). The W3C VC data model ecosystem follows 4 basic roles that are followed as shown in Fig. 1 [23].

Firstly, an issuer is responsible for generating & issuing verifiable credentials about a subject. Secondly, a verifier computes the cryptographic proof to verify the legitimacy and authenticity of credentials about a particular subject. Thirdly, a holder receives and manage verifiable credentials from issuers, and create verifiable presentations which are presented to verifiers as proofs. A verifiable data registry maintains identifiers and schemas. The conceptual architecture is portrayed in Fig. 1.

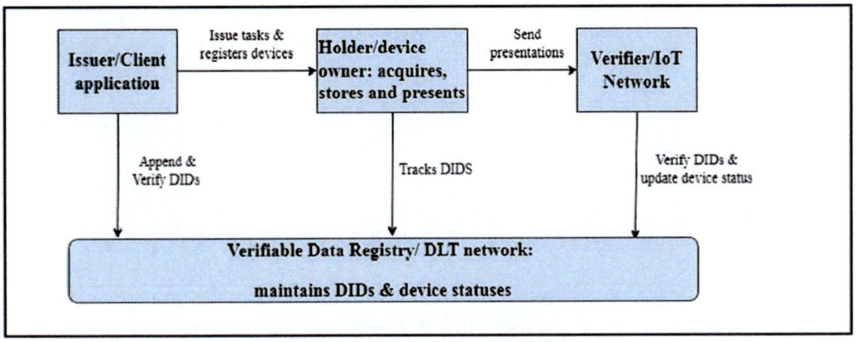

Fig. 1. Conceptual architecture.

This solution aims to ensure that device enrolment is secure, thus, prior to any IoT device being permitted to interact with other devices, it must be enrolled first following an enrolment process. during this process, each device is provided with a unique credential. To register, the device owner submits the registration request to the system. This request is handled by the client app which issues a secret for the enrolment process through the client app. For this process, public and private keys are used along with the unique DID. A DID Document is then generated and is binded with the public key. the timestamp and DID is then Sent to the DLT for timestamping purposes. The DID is then sent to the device owner to be used for authentication purpose shown in Fig. 2. After the enrolment process, the device owner is allowed to access and consume services provided by the network. To use the network a device must first be authenticated. To achieve this, a DID authentication challenge in a form of JSON Web Token (JWT) that is authenticated by the User's JSON Web signature. The user signs the challenge using their associated private key and sends it back so that access can be verified. If verified, access is then granted, and this process is depicted in Fig. 2.

With this solution, the device owner has control over all the devices and to whom they are to communicate with. Thus, improving security and ensuring privacy. With this approach, we can achieve some of the self-sovereign identity requirements namely, control, access, existence, transparency, etc. by Christopher Allen [24]. It is worth noting that, this is a proposed solution for ensuring that IoT devices are secure using DLTs and W3C standards. Therefore, performance metrics can only be shared at a later stage once the solution has been implemented and tested accordingly.

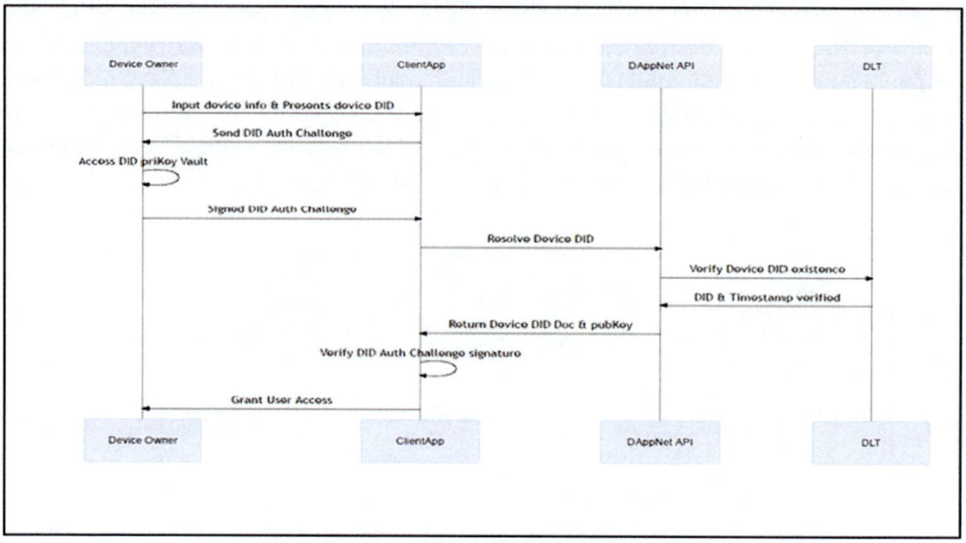

Fig. 2. Device authentication flow diagram.

5 Conclusion and Future Work

In this paper, we have carried out an in-depth systematic review about the integration of DLT with of IoT to ensure secure IoT devices. It can be observed that the IoT scale is growing tremendously, and more researchers are investing efforts towards finding ways to integrate these two newly introduced technologies that are takin the world by storm (DLT and IoT). However, this growth can affect the performance of the DLT technology when integrated without performing an evaluation. For example, the POW consensus protocol involves many issues in the blockchain-IoT environment including throughput, delay, and high computing resource consumption, which affect the efficiency of IoT management [25–27]. Hence, it is essential to consider the proper consensus protocol and the resource constraints introduced by IoT devices when implementing the environment. In order to successfully implement a secure blockchain-IoT environment and communication protocols, energy and storage requirements or resource-constrained devices should be considered and managed. Hence, designing lightweight communication protocols or implementing energy managements techniques are challenging issues in a blockchain-based IoT environment. Thus, the importance of this research.

For this research, a background around IoT and DLTs have been shared as well as what has been done in terms of integrating these two platforms thus far. This paper presented a proposed solution for the integration of DLTs and IoT whilst being cognisant of the challenges presented by IoT devices. The solution employed 4 techniques: W3C standards, VC, DIDs and DLTs. The standards are still new and emerging but will be vital in order to have a standardise protocol to follow for the integration process are still a few changes and evolution being implemented for these standards, but they have promising implementations that require.

These standards are still being changed on a yearly basis, whereby new improvements are introduced to improve the performance and their adoption as well. For standardisation processes, it is vital to find and adopt standards that could ensure that the integration process is smooth and can be internationally recognised. The proposed solution goes further into portraying the flow of information during different stages of using the system i.e. When a user wants to enrol or register a device on the network, device-to-device communication, etc. As this is still a new field of research, more efforts are still required to ensure that it is properly integrated and can be standardised. Hence, further research efforts are still required in the near future to make the integration of DLTs a permanent and mature approach in the context of IoT device environment as well as ensuring security.

With the existing research, it becomes evident that more researchers are interested in this topic. Further research and experimentations are still required for the assessment of all possible DLTs to be used for this solution from the 3 types of DLTs discussed in this paper. From there, the proposed solution can then be tested. From the assessment, it can be important to develop or implement IoT specific DLT testbench using the cheapest and smallest possible embedded devices to demonstrate the effectiveness and practicality.

References

1. Bouras, M.A., Lu, Q., Dhelim, S., Ning, H.: A lightweight blockchain-based IoT identity management approach. Future Internet **13**(2), 24 (2021). https://doi.org/10.3390/fi13020024
2. Pavithran, D., Shaalan, K., Al-Karaki, J.N., Gawanmeh, A.: Towards building a blockchain framework for IoT. Clust. Comput. **23**(3), 2089–2103 (2020). https://doi.org/10.1007/s10586-020-03059-5
3. Abed, S., Jaffal, R., Mohd, B.J., Al-Shayeji, M.: An analysis and evaluation of lightweight hash functions for blockchain-based IoT devices. Cluster Comput. **24**(4), 3065–3084 (2021). https://doi.org/10.1007/s10586-021-03324-1
4. Tschirner, S., Zeuch, K., Kaven, S., Bornholdt, L., Skwarek, V.: Security in Distributed Systems by Verifiable Location-Based Identities (2023). arXiv preprint arXiv:2302.14713
5. Pretorius, M., Mthethwa, S.N.: A survey of distributed trust mechanisms suitable for IoT devices (2019)
6. Javaid, M., Khan, I.H.: Internet of Things (IoT) enabled healthcare helps to take the challenges of COVID-19 Pandemic. J. Oral Biol. Craniofacial Res. **11**(2), 209–214 (2021). https://doi.org/10.1016/j.jobcr.2021.01.015
7. King, J., Awad, A.I.: A distributed security mechanism for resource-constrained IoT devices. Informatica **40**(1), 1–12 (2016)
8. Mahmoud, R., Yousuf, T., Aloul, F., Zualkernan, I.: Internet of things (IoT) security: current status, challenges and prospective measures. In: 2015 10th international conference for internet technology and secured transactions (ICITST), pp. 336–341. IEEE (2015)
9. Sadique, K.M., Rahmani, R., Johannesson, P.: Identity management in internet of things: a software-defined networking approach. In: Kundu, S., Acharya, U.S., De, C.K., Mukherjee, S. (eds.) Proceedings of the 2nd International Conference on Communication, Devices and Computing. LNEE, vol. 602, pp. 495–504. Springer, Singapore (2020). https://doi.org/10.1007/978-981-15-0829-5_48
10. Fan, X., Chai, Q.: Roll-DPos: a randomized delegated proof of stake scheme for scalable blockchain-based Internet of Things systems. In: ACM International Conference Proceeding Series, pp. 482–484 (2018). https://doi.org/10.1145/3286978.3287023

11. Nakamoto, S.: Bitcoin: a peer-to-peer electronic cash system (2008). www.bitcoin.org/bitcoin.pdf
12. Alaba, F.A., Othman, M., Hashem, I.A.T., Alotaibi, F.: Internet of Things security: a survey. J. Netw. Comput. Appl. **88**, 10–28 (2017)
13. Saxena, S., Bhushan, B., Ahad, M.A.: Blockchain based solutions to secure IoT: background, integration trends and a way forward. J. Netw. Comput. Appl. **181**, 103050 (2021). https://doi.org/10.1016/j.jnca.2021.103050
14. Atlam, H.F., Wills, G.B.: Intersections between IoT and distributed ledger. In: Advances in Computers, 1st edn, vol. 115, no. January. Elsevier Inc. (2019). https://doi.org/10.1016/bs.adcom.2018.12.001
15. Farahani, B., Firouzi, F., Luecking, M.: The convergence of IoT and distributed ledger technologies (DLT): opportunities, challenges, and solutions. J. Netw. Comput. Appl. **177**, 102936 (2021). https://doi.org/10.1016/j.jnca.2020.102936
16. Reyna, A., Martín, C., Chen, J., Soler, E., Díaz, M.: On blockchain and its integration with IoT. challenges and opportunities. Future Gener. Comput. Syst. **88**, 173–190 (2018). https://doi.org/10.1016/j.future.2018.05.046
17. Issa, W., Moustafa, N., Turnbull, B., Sohrabi, N., Tari, Z.: Blockchain-based federated learning for securing internet of things: a comprehensive survey. ACM Comput. Surv. **55**(9), 1–43 (2023)
18. Ferrag, M.A., Derdour, M., Mukherjee, M., Derhab, A., Maglaras, L., Janicke, H.: Blockchain technologies for the internet of things: research issues and challenges. IEEE Internet Things J. **6**(2), 2188–2204 (2018)
19. Panikkar, S., Nair, S., Brody, P., Pureswaran, V.: ADEPT: an IoT practitioner perspective (2015). https://ibm.biz/devicedemocracy
20. Majer, A.: Slock.IT: enabling IoT and the universal sharing network. In: Blockchain Research Institute (BRI), December, pp. 1–18 (2017)
21. Rejeb, A., Keogh, J.G., Treiblmaier, H.: Leveraging the internet of things and blockchain technology in supply chain management. Future Internet **11**(7), 161 (2019)
22. Yin, J., et al.: SmartDID: a novel privacy-preserving identity based on blockchain for IoT. IEEE Internet Things J. **10**, 6718–6732 (2022)
23. Sporny, M., Longley, D., Chadwick, D.: Verifiable credentials data model 1.0: expressing verifiable information on the web. In: W3C Recommendation, 19 November 2019 (2019). https://www.w3.org/TR/vc-data-model/
24. Allen, C.: The path to self-sovereign identity. Life With Alacrity (2016). http://www.lifewithalacrity.com/2016/04/the-path-to-self-soverereign-identity.html
25. Tewari, A., Gupta, B.B.: Security, privacy and trust of different layers in Internet-of-Things (IoTs) framework. Future Gener. Comput. Syst. **108**, 909–920 (2020). https://doi.org/10.1016/j.future.2018.04.027
26. Lockl, J., Schlatt, V., Schweizer, A., Urbach, N., Harth, N.: Toward trust in internet of things ecosystems: design principles for blockchain-based IoT applications. IEEE Trans. Eng. Manag. **67**(4), 1256–1270 (2020). https://doi.org/10.1109/TEM.2020.2978014
27. Mistry, I., Tanwar, S., Tyagi, S., Kumar, N.: Blockchain for 5G-enabled IoT for industrial automation: a systematic review, solutions, and challenges. Mech. Syst. Signal Process. **135**, 106382 (2020). https://doi.org/10.1016/j.ymssp.2019.106382

Check for updates

Analysis and Forecast of Energy Demand in Senegal with a SARIMA Model and an LSTM Neural Network

Moustapha Drame[1], Djamal Abdoul Nasser Seck[1(✉)], and Baye Samba Ndiaye[2]

[1] Faculty of Science and Technology, Department of Mathematics-Computer Science, Cheikh Anta Diop University of Dakar (UCAD), Dakar, Senegal
{moustapha4.drame,djamal.seck}@ucad.edu.sn

[2] Department of General Economics Studies, SENELEC, Energy Demand Management Service, Dakar, Senegal
bayesamba.ndiaye@senelec.sn

Abstract. In this paper, we propose models to forecast the demand of energy to allow Senelec to be able to maintain the balance between supply and demand of electricity in Senegal. We first analyze the consumption behavior of the different categories of customers accordingto Senelec's 2020 billing database. After we present two prediction models of the energy demand of Senelec's MediumVoltage customers: a SARIMA model based on classical statistical modeling of time series and an LSTM neural network model based on artificial intelligence. The out-of-sample predictive performance of the LSTM model is globally better than that of the SARIMA model according to the indicators (MAE and RMSE (RMSE: Roote Mean Squart Error)) we used to compare them. For the LSTM model, we have a Mean Absolute Error (MAE (MAE: Mean Absolute Error)) of 0.148597 on the test data while, for the SARIMA model, we have a Mean Absolute Error (MAE) of 0.166997. These two models have been integrated into a web application to offer a simple and user-friendly interface for the prediction of energy demand in Senegal.

Keywords: Energy · demand · production · time series · model · ARMA · SARIMA · SARIMAX · Neural Networks · Recurrent Neural Networks (RNN) · LSTM

1 Introduction

Electricity is a fundamental factor of production in economic activity. Between 2000 and 2011, Senegal faced a deep energy crisis with an almost permanent intermittency and high cost of electricity [1]. But in recent years there has been a great improvement in the supply of electricity with the increase of the capacity of the production park from 851 MW to 1484 MW between 2013 and 2020. Despite this improvement, the regularity of electricity supply remains a particular challenge. Not only is electricity a continuous flow that is difficult to store,

M. Younas et al. (Eds.): Deep-BDB 2023, LNNS 768, pp. 129–140, 2023.
https://doi.org/10.1007/978-3-031-42317-8_11

but its demand is also subject to significant fluctuations because of its dependence on variables related to the country's economic activities, including GDP. Therefore, it is necessary to adopt a holistic approach to define a coherent and solid vision for good control of the energy sector. Senelec, which is the structure in charge of the electricity sector in Senegal, is studying strategies such as forecasting electricity demand for efficient management of resources allocated to production. Indeed, forecasting electricity demand makes it possible to set up decision support tools and minimize the costs of over-production or under-production. But forecasting errors should be avoided as they can cause significant operational costs [2]. In 1984, an increase of 1% in the forecasting error in an electrical installation in the United Kingdom caused an annual increase in operating costs of£10 million [4]. On the other hand, a reduction of 1% in the average forecast error could save thousands or even millions of dollars in an electricity production unit [3]. In this article we will first analyze Senelec's 2020 billing database. Then we will propose a model for forecasting the electricity demand of Medium Voltage customers (also called special customers) who are the main actors in the economic activity of Senegal.

2 Analysis of Woyofal Customers Database

The database of Woyofal customers (prepayment customers) is made up of the following data: transactions, the amount purchased in FCFA and the energy in kwh of customers. In this database we have four categories of customers: Domestic Small Power (DPP), Domestic Medium Power (DMP), Professional Small Power (PPP) and Professional Medium Power (PMP). Figure 1 shows the distribution of the Woyofal customers by category. It can be seen that DPP category represents 78% of woyofal customers, PPP category represents 20% and DMP and PMP categories together represent 2%.

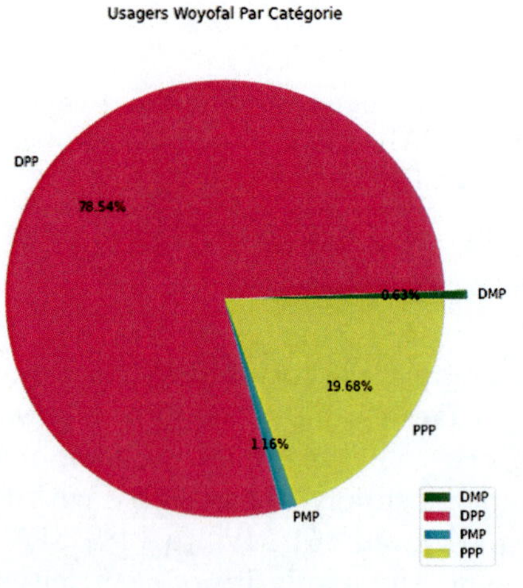

Fig. 1. Distribution of the Woyofal customers by category.

Using the K-Means clustering method, the classification of a sample of DPP customers (the majority of Woyofal customers) into three classes according to the average purchase and the frequency of purchase gives the following results:

- a class of 552175 customers who buy an average of 3,467 FCFA every 20 days.
- a class of 76691 customers who buy an average of 13,339 FCFA every 21 days.
- a class of 4540 customers who buy an average of 54,792 FCFA every 25 days.

Figure 2 is an illustration of this classification of DPP customers.

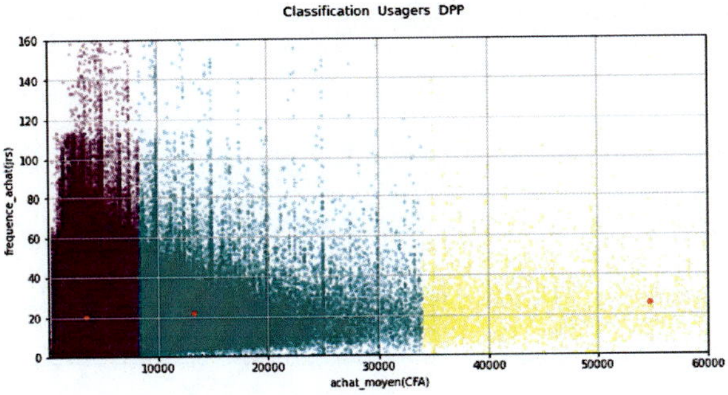

Fig. 2. classification of DPP customers

3 Building a Forecasting Model of Electricity Demand

In this section, we propose a model for forecasting the electricity demand of Medium Voltage (MV) customers who are the main actors in economic activity in Senegal. Medium voltage (MV) is characterized by: the provision of a transformer station with a subscribed power of more than 34 kW, energy billing according to consumption hours, a peak hour from 7 p.m. to 11 p.m. and an off-peak hour from 11 p.m. to 7 p.m. We propose two types of models for forecasting the electricity demand of MV customers. A SARIMA model based on statistical methods andan LSTM neural network based on machine Learning which is a branch of artificial intelligence. We use time series data to create the forecast model. The first time series is the monthly consumption of MV customers from 2010 to 2021. The second time series is the evolution of GDP in Senegal and represents the exogenous variable that influences the electricity consumption of MV customers.

3.1 Forecasting Electricity Demand with a SARIMA Model

This method of modeling time series is based on the stationarity of the series. If a series is stationary and exhibits a particular behavior during a given interval of time, we can safely assume that it will present the same behavior at a later time. The SARIMA or seasonal ARIMA model is an extension of the ARIMA model, which all belong to ARMA processes. The ARMA process is a stationary process composed of two terms, an autoregressive term (AR) and a moving average (MA).

The auto-regressive process AR(p) works on the assumption that past values have an effect on current values, given by the following formula:

$$y_t = a_1 y_{t-1} + a_2 y_{t-2} + a_3 y_{t-3} + \ldots + a_p y_{t-p} \tag{1}$$

The Moving Average MA(q) is the following random process:

$$X_t = \varepsilon_t + b_1 \varepsilon_{t-1} + b_2 \varepsilon_{t-2} + b_3 \varepsilon_{t-3} + \ldots + b_q \varepsilon_{t-q} \tag{2}$$

The combination of these two processes (AR(p) and MA(q)) gives the ARMA(p,q) process:

$$Y_t = a_1 y_{t-1} + a_2 y_{t-2} + \ldots + a_p y_{t-p} + \varepsilon_t + b_1 \varepsilon_{t-1} + b_2 \varepsilon_{t-2} + \ldots + b_q \varepsilon_{t-q} \tag{3}$$

The autoregressive moving average (ARMA) model requires the time series to be stationary to make predictions. There are multiple transformations that can be applied to the series to make it stationary. Differentiation is one of the most used methods to eliminate trends in a time series. It consists in subtracting the current value from the previous one.

First-order differentiation eliminates linear tendencies by the following transformation:

$$z_i = (y_i - y_{i-1}) \tag{4}$$

Second-order differentiation deals with quadratic trends by the following transformation:

$$z_i = (y_i - y_{i-1}) - (y_{i-1} - y_{i-2}) \tag{5}$$

This differentiation integrated into the ARMA process gives the ARIMA(p,d,q) model. The SARIMA model has the same parameters (p, d, q) as ARIMA in addition to the parameters (sp, sd, sq, s):

- **s:** Period of seasonality
- **sp:** Seasonal auto-correlation
- **sd:** Derivation of seasonality
- **sq:** Seasonal moving average

There is a version of SARIMA called SARIMAX that takes into account exogenous variables to model a time series. We will use it to forecast the electricity demand of MV customers with GDP as the exogenous variable. Following are some examples of the time series data between 2010 and 2021 that we use to build the model.

date	Energy(MWH)	PIB
2010-01	55763.623	692.596221
2010-02	39013.692	693.999053
2010-03	49126.950	695.404726
2010-04	47354.638	696.463751
2010-05	52229.414	697.699269
⋮	⋮	⋮
2021-11	100540.689	1240.987187

Figure 3 represents the evolution over time of the electricity consumption of MV customers. It shows that consumption increases on average every year. We can notice an irregular fluctuation in consumption between 2019 and 2021, which can be explained by the covid19 health crisis. To take into account this type of phenomenon we add in the model random terms or white noise.

Figure 4 gives the decomposition of the time series into three main components:

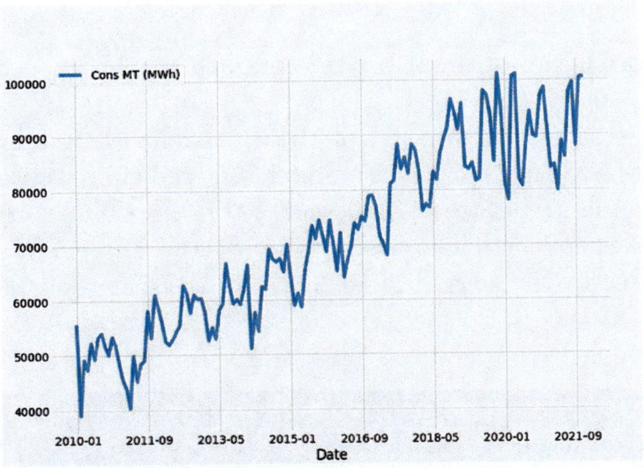

Fig. 3. MV customers consumption

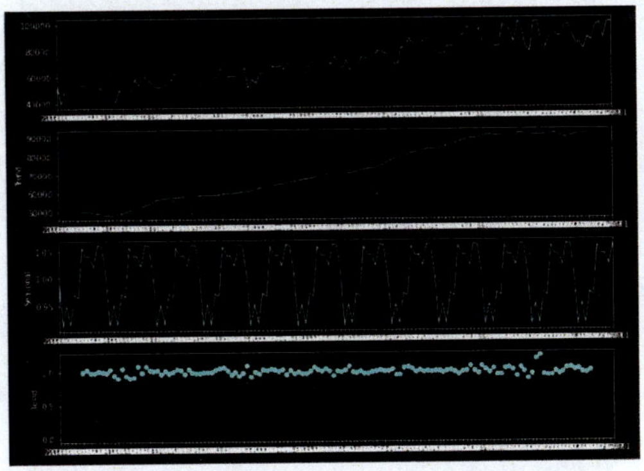

Fig. 4. Decoposition of the time series

- **Trend:** shows that the series has an increasing trend over time
- **Seasonnal:** shows that the series has a seasonal component, for example consumption in March is identical for each year because of the heat.
- **Resid:** these are random phenomena that are added to this periodicity.

The optimal parameters of the SARIMA model are those which minimize the Akaike information criterion (AIC) given by:

$$AIC = 2k - 2\ln(L)$$

where k is the number of parameters to be estimated from the model and L is the maximum likelihood function of the model. The python pmdarima library contains the auto-arima function to perform a step-by-step search for the parameters which give the lowest value of AIC. After estimating the parameters with the auto-arima function, we have the following result:

```
Best model:  ARIMA(1,1,1)(1,0,1)[12]
Total fit time: 50.053 seconds
```

In other words, the optimal parameters are $p = 1$, $q = 1$, $d = 1$, $sp = 1$, $sq = 1$ and $sd = 0$.

Figure 5 illustrates the forecast of the electricity demand of MV customers with our SARIMAX(1,1,1)(1,0,1,12) model. The red curve represents the forecast of electricity demand between 2010 and 2023. The blue curve represents the actual values of the electricity consumption of MV customers. The convergence of the two curves in the period 2010–2021 proves the good performance of the model.

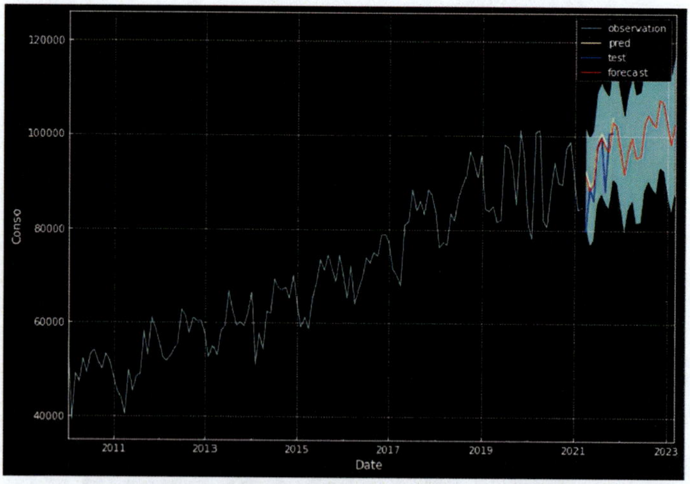

Fig. 5. Consumption forecast for 2023

3.2 Forecasting Electricity Demand with an LSTM Neural Network

Long Short Term Memory (LSTM) Neural network [7] is a special type of Artificial Neural Network used in machine learning to process sequential data. Machine Learning (ML) is a branch of Artificial Intelligence that enables machines to learn patterns from data using automated methods.

Given a set of input-output data, a supervised Machine Learning method can learn the mappings between the inputs and the outputs and, then, predict the output for new input data. In other words, it makes it possible to find in the data a function φ such that:

$$Y = \varphi\left(X_1, X_2, \ldots, X_p\right) \tag{6}$$

where $X1, X2, ..., XP$ are the independent variables representing the input data and Y is the dependent variable representing the output data.

The function φ is a model that enables to predict Y for a new input data w given its attributes $X1(w), X2(w), ..., XP(w)$. It is a regression model if Y is a quantitative variable and it is a classification model if Y is a qualitative variable.

An artificial neural network is a machine-learning model that is inspired by the functioning of the human brain. It is composed of layers of artificial neurons with weighted connections between neurons. An artificial neuron is a computational unit that receives values xj through its input connections, adds a value called activation threshold or bias b, and applies an activation function to compute an output y as illustrated by Fig. 6.

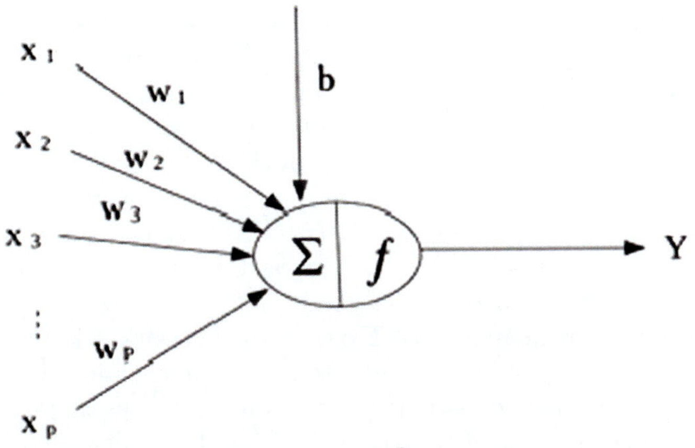

$$Y = f\left(\sum_{j=1}^{p} w_j X_j + b\right)$$

Fig. 6. Artificial neurone

The Multilayer Perceptron [8] is a type of artificial neural network used in supervised learning for classification or regression. It is organized in several layers, each consisting of a variable number of neurons: an input layer, one or more hidden layers and an output layer. Each neuron in a layer is connected to all neurons in the next layer and there are no connections between neurons in the same layer. The connections between neurons are characterized by weights which are real numbers Fig. 7 shows an example of a multilayer perceptron.

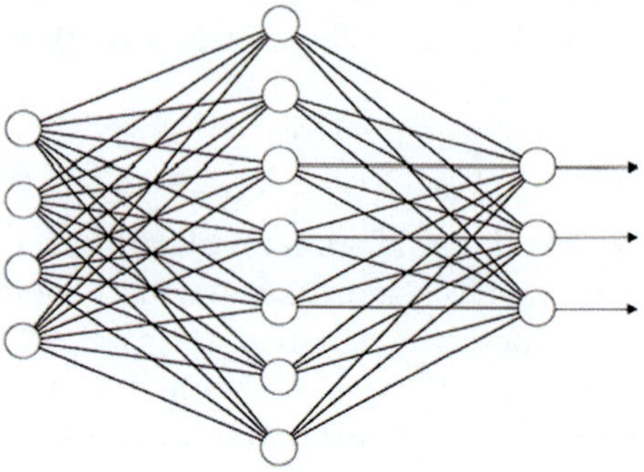

Fig. 7. Multilayer Perceptron

The multilayer perceptron is a feed-forward neural network, which means that information flows from the input layer to the output layer. The neurons of the input layer receive the input data and send them to the neurons of the first hidden layer which apply treatments to them with their activation function and then, in turn, send their outputs to the neurons of the next layer, and so on until the neurons of the output layer. The outputs of the neurons of this last layer represent the prediction of the neural network.

Supervised Learning of the Multilayer perceptron consists of iteratively modifying the weights of the connections between the neurons in order to minimize the prediction error. It is performed by the back-propagation algorithm [8,9] and the gradient descent optimization method.

A long short-term memory (LSTM) neural network [7] is a special type of neural network that is particularly adapted to sequential data because it is capable of learning long-term dependencies between time steps of data. In contrast to the MLP Feedforward Neural Network, the LSTM neural network is a recurrent neural network, which contains feedback connections in the hidden layers whose neurons are replaced by units called memory blocks. Each unit is composed of a cell, which is the memory part of the unit, and three gates: an input gate, an output gate and a forget gate. Figure 8 shows a schematic description of an LSTM unit.

An LSTM unit can remember values over arbitrary time steps. The output from the previous time step, h_{t-1}, and the current input x_t, are the input to the block at time t, then the output h_t at time t will be an input to the same

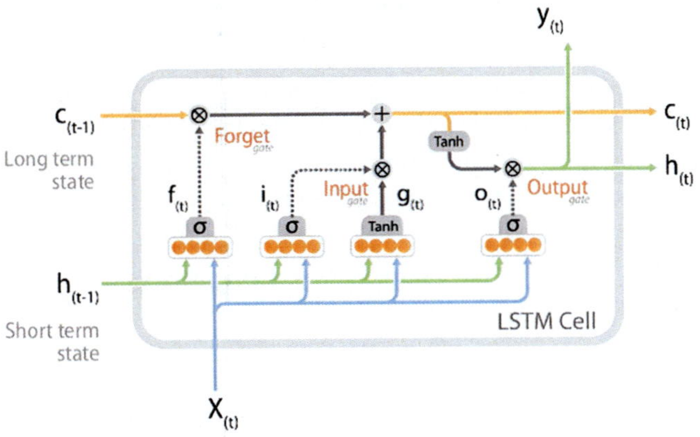

Fig. 8. An LSTM unit

block in the next time step $t+1$. The three gates control the flow of information through the cell.

The input gate it receives the current input x_t and the output of the previous time h_{t-1} and applies the sigmoid activation function to return values between 0 and 1. A value of 0 means that the information is blocked, whereas a value of 1 means that the information pass.

$$i_t = \sigma\left(W_{xi}X_t + W_{hi}h_{t-1} + b_i\right)$$

The cell input layer has a similar flow as the input gate, receiving $ht-1$ and x_t as input. However, a tanh activation is used to return values between -1 and 1.

$$g_t = \tanh\left(W_{xg}X_t + W_{hg}h_{t-1} + b_g\right)$$

The forget gate f_t determines which information from previous steps of the cell should be memorized or forgotten. It applies the sigmoid function to return a value between 0 and 1 based on the input, h_{t-1} and x_t. When this value is close to 0, the old state does not matter anymore.

$$f_t = \sigma\left(W_{xf}X_t + W_{hf}h_t + b_f\right)$$

The cell state stores the memory of a cell over a longer time period. At time t, the current cell state c_t is updated by the previous cell state c_{t-1} controlled by the forget gate and the product of the current input and the cell input.

$$c_t = f_t \otimes c_{t-1} + i_t \otimes g_t$$

The output gate o_t applies the sigmoid activation function to control the flow of information out of the cell. The values of the output gate at time t allow to control the current cell state ct activated by a tanh function to obtain the final output vector y_t which is duplicated and goes into the short memory h_t

$$o_t = \sigma\left(W_{xo}X_t + W_{ho}h_{t-1} + b_o\right)$$

$$y_t = h_t = o_t \otimes \tanh(c_t)$$

This type of operation of the LSTM neural network proves its potential to learn long-term and short-term dependency relationships and that it is suitable for time series analysis. So we propose to train an LSTM model with our data to forecast the electricity demand of MV customers. The Keras and Tensorflow libraries allow to train and optimize our LSTM model.

Figure 9 shows evolution of the training process. As we can notice, the error of the model decreases at each epoch thanks to the gradient descent optimization algorithm.

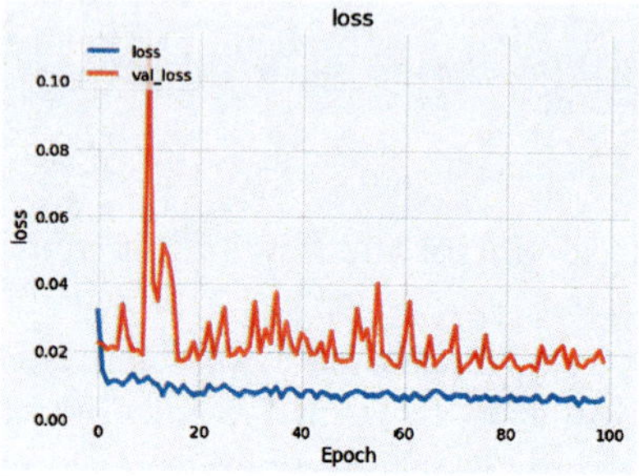

(a) loss fonction.

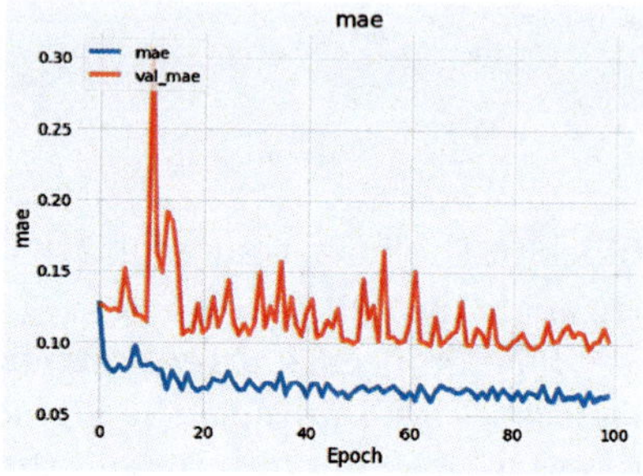

(b) Mean Absolute Error

Fig. 9. LSTM Learning Curve

Figure 10 shows the comparison between the forecast test data and the actual data as well as the data from the past.

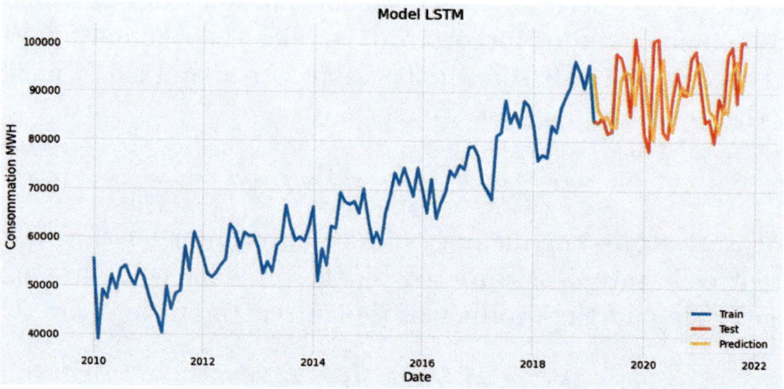

Fig. 10. Consumption forecast with (LSTM)

Using the performance indicators which are the RMSE and the MAE, our LSTM model makes an RMSE error of 0.01156 and an MAE error of 0.148597 on the test data.

4 Deploying the Forecasting Model in a Web Application

In this section, we present the web application developed in python with the Django framework to deploy our SARIMA and LSTM prediction models and offer a user-friendly interface for predicting energy demand at a given date. Figure 11 shows the interface of the application with a red curve and a green curve showing the energy demands predicted respectively by the LSTM model and the SARIMA model for different dates entered successively. The blue curve shows the actual values of energy demand on the same dates and makes it possible to visually compare the predictive performance of the two models.

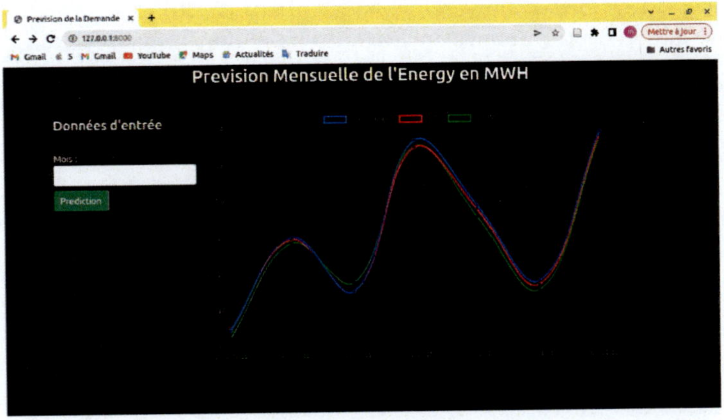

Fig. 11. web application for electricity demand forecast

5 Conclusion

In this article, after the analysis of Senelec's 2020 billing database, we have proposed a SARIMA model and an LSTM Neural Network model to forecast the electricity demand of Medium Voltage customers. The out-of-sample predictive performance of the LSTM model is globally better than that of the SARIMAX model according to the indicators we used to compare them, namely the MAE and RMSE. Thus for the SARIMA model, we have an MAE of 0.166997 on the test data and for the LSTM model, we have an MAE of 0.148597.

As a perpective, the web application we have developped to deploy the models can be improved by linking it to electricity consumption data collected in real-time with smart meters.

References

1. SMES, Senegal Energy Management Strategy (SMES) (2021)
2. Haida, T., Muto, S.: Regression based peak load forecasting using a transformation technical. IEEE Trans. Power Syst. **9**, 1788–1794 (1994)
3. Hobbs, B., Jitprapaikulsarn, S., Konda, S., Chankong, V., Loparo, K., Maratukulam, D.: Analysis of the value for unit commitment of improved load forecasts. IEEE Trans. Power Syst. **14**, 1342–1348 (1999)
4. Soares, L.J.: Medeiros, MC modeling and forecasting short-term electricity load?: a comparison of methods with an application to Brazilian data. Int. J. Forecast. **24**, 630–644 (2008)
5. Bellahsen, A.: Artificial intelligence for the optimization of electrical energy in a smart grid. PhD Thesis, Polytechnique Montreal (2020)
6. Talbi, R.: Le preceptron multicouches approximateur universel. Ph.D. thesis, Alger (2005)
7. Hochreiter, S., Schmidhuber, J.: Long short-term memory. Neural Comput. **9**(8), 1735–1780 (1997)
8. Werbos, P.J.: Beyond Regression: New Tools for Prediction and Analysis in the Behavioral Sciences. Doctoral Dissertation, Harvard University, Cambridge (1974)
9. Rumelhart, D.E., Hinton, G.E., Williams, R.J.: Learning representations by back-propagating errors. Nature **323**, 533–536 (1986)

Author Index

M. Younas et al. (Eds.): Deep-BDB 2023, LNNS 768, p. 141, 2023.
https://doi.org/10.1007/978-3-031-42317-8